BEYOND THE SEA

BEYOND THE SEA

THE HIDDEN LIFE
in LAKES, STREAMS,
and WETLANDS

DAVID STRAYER

Johns Hopkins University Press
Baltimore

2 4 6 8 9 7 5 3 1

Johns Hopkins University Press

2715 North Charles Street

Baltimore, Maryland 21218

www.press.jhu.edu

Library of Congress Cataloging-in-Publication Data

Names: Strayer, David Lowell, 1955– author.
Title: Beyond the sea : the hidden life in lakes, streams,
and wetlands / David Strayer.
Description: Baltimore : Johns Hopkins University Press,
2024. | Includes bibliographical references and index.
Identifiers: LCCN 2024010575 | ISBN 9781421450070
(hardcover) | ISBN 9781421450087 (ebook)
Subjects: LCSH: Freshwater ecology. | Freshwater animals. |
Freshwater biodiversity. | Freshwater biodiversity conservation.
Classification: LCC QH541.5.F7 S77 2024 | DDC 577.6—dc23/eng/20240620
LC record available at https://lccn.loc.gov/2024010575

A catalog record for this book is available from the British Library.

Special discounts are available for bulk purchases of this book. For more
information, please contact Special Sales at specialsales@jh.edu.

CONTENTS

CHAPTER 14

Solutions: Protecting and Restoring Inland-Water Ecosystems

154

CHAPTER 15

Back to the Theme: Closing Remarks

168

though I have looked everywhere
I can find nothing lowly
in the universe

—A. R. AMMONS, "STILL"

PREFACE

As a scientist who has worked on freshwater ecology for most of my life, I am familiar with the enormous diversity of inland waters and the life that they contain, and I worry about the deadly peril that these ecosystems and their inhabitants face from human activities. Inland-water ecosystems contain far more than their share of biodiversity based on the tiny area of Earth that they occupy and are far more imperiled than marine or terrestrial ecosystems. These matters are well established and are well known among the scientists who study inland waters.

When I speak to nonscientists, though, whether on a river bank or in an auditorium, I am often struck by the disconnect between what scientists know and what regular people know about inland waters. Many people are familiar with Earth's most famous biodiverse ecosystems, such as coral reefs and tropical rainforests. They understand that these ecosystems are endangered by human activities and may even be engaged in efforts to protect them. But few are familiar with the basic facts about inland-water biodiversity or how humans have imperiled so many inland-water species.

This book introduces you to a few of the most remarkable inland waters and their amazing inhabitants and to some of the perils that they face. I show you that inland waters include some of the most varied ecosystems on Earth, that they are exceptionally biodiverse and contain species with fascinating adaptations and ways of life, and that they have been badly damaged by human activities and require careful management if we are avoid future damage. I hope that this book imparts some sense of wonder about inland-water ecosystems, sadness about the disappearance of so many inland-water species, and passion about protecting inland-water biodiversity for the future.

Instead of trying to offer a comprehensive treatment of inland-water ecology and biodiversity, I try to make my points by offering just a few select examples of the wide range of inland waters and the life that they contain. This approach lets me keep the book short, avoid having to include highly technical material (hardly anyone enjoys differential equations, in my experience), and choose the most engaging examples, concentrating on the unfamiliar, the extreme, and the unbelievable. You shouldn't need any specialized knowledge or training in biology, ecology, or limnology (or even have to know what "limnology" means) to read this book.

The first part of the book (chapters 1–5) describes some of the great variety in the physical characteristics and life stories of different inland waters, which come in all shapes and sizes and have all kinds of histories. Chapter 6 is a brief (and I hope painless) introduction to the chemical diversity of inland waters, and chapter 7 introduces the idea that inland waters are basically islands, a concept that is of paramount importance in understanding the evolution and conservation of inland-water life. Chapter 8 briefly describes the kinds of species that live in inland waters and provides statistics about the extraordinary biological richness of these habitats. But the real biology is contained in chapters 9 to 12, which discuss adaptations that inland-water inhabitants possess that help to deal with four selected challenges: keeping from getting washed down to the sea, managing when the water dries up, finding food, and (of course) sex. The book concludes with a depressing chapter 13, which describes the heavy human impacts on inland waters and the poor state of inland-water biodiversity, and a slightly less depressing chapter 14, which outlines some of the ways that we can protect and restore inland-water ecosystems.

This book wouldn't exist without the help and encouragement of friends and colleagues. Josh Ginsberg, Gene Likens, and Bill Schlesinger, my bosses at the Cary Institute, all encouraged me to write for and speak to the public. The Graham Sustainability Institute at the University of Michigan provided a cordial academic home for the past few years. Many people encouraged me over the years by critiquing drafts or offering kind words; I'd especially like to recognize Judy Bondus, Alan Berkowitz, Jill Cadwallader, Jon Cole, Lisa Dellwo, Pam Freeman, Alex Haddad, Steve Hamilton, Lia Harris, Dave Poulson, Lori Quillen, Jen Read, Bill Schlesinger, Amy Schuler, Dan Shapley, Leslie Tumblety, John Waldman, Joe Warner, and many other friends at the Cary Institute. I thank the many colleagues who shared information and stories with me and especially the photographers who posted their beautiful photographs for general use or agreed to let me include them. I am very grateful to David Dudgeon, Chris Solomon, and an anonymous reviewer who read a draft of the entire manuscript and offered many helpful suggestions, to Caryn Vaughn for her comments on the proposal for this book, and to MJ Devaney for her keen editorial eye.

BEYOND THE SEA

Chapter 1

THEME AND VARIATIONS
The Rest of the Blue Planet

When we refer to Earth as the "blue planet," we are thinking of our marvelous ocean. The famous blue marble photograph of Earth that NASA took from outer space is blue because of the ocean that covers 71% of our planet.[1] Not only does the ocean cover most of our watery planet but it also contains remarkable ecosystems and species. Most of us know about these highly varied and fascinating ecosystems—the clear, blue waters of the open ocean, inky depths miles beneath the waves, coral reefs, deep-sea vents, kelp forests, rocky shores and sandy beaches, bright aqua tropical flats, ice-covered polar seas, salt marshes, and so on. And most of us know about the stunning diversity of ocean life, ranging from singing whales and huge squids to deep-sea fishes that carry their own flashlights and the colorful life of the coral reef, knitted together into complex, interconnected communities. Earth's ocean really is wonderful.

But there are other blue bits on our planet far less famous and far smaller than the ocean, indeed so small that they are scarcely even visible in the blue marble photograph. These bits are what we call "inland waters," and they include lakes, ponds, streams, wetlands, and groundwaters, which together cover only a tiny fraction of Earth.

Although we depend on these inland waters every day for drinking and irrigation water, transportation, power production, food, waste disposal, recreation, and a hundred other things, they don't occupy the same place in the public imagination as the ocean. Children dream of growing up to become marine biologists and oceanographers, not someone who explores the pond down the street. There is no Jacques Cousteau of the creeks, no Sylvia Earle of the woodland pools.

Yet in their own modest way, inland waters are just as marvelous as the ocean. They are extraordinarily varied in terms of nearly every aspect of their physical structure and chemistry, and the range of environmental conditions they present in many ways surpasses that of the ocean. Partly because of this great variation, inland waters support a biological diversity all out of proportion to the small area that they cover.[2] About 10% of all known species on Earth and a half of all known fish species live in inland waters.[3] As in the case of ocean species, many of these species are beautiful, fascinating, or valuable to people, and worthy of our attention. Because humans use (and abuse) inland waters so heavily (and often so thoughtlessly), they have damaged or destroyed many inland-water habitats and killed their inhabitants. As a result, the proportion of species that are endangered is very much higher in inland waters than on land or in the ocean. Many inland-water species are already extinct, among which were some of the world's most intriguing animals, and perhaps tens of thousands of additional species are so imperiled that they may not live to see the year 2100.

This book's main goal is to show you some of the beauty and wonder of inland waters and the life they contain. I hope to convince you that the modest inland waters that constitute the rest of the blue planet are marvelous, just like the ocean. I also describe how human activities have damaged inland-water ecosystems, jeopardizing many of the species that live there, and outline a few actions that we can take, individually and collectively, to protect and restore these ecosystems.

This book is not a comprehensive textbook of inland-water ecology or even a comprehensive listing of all of the cool stuff that scientists know about inland waters. Instead, it offers a selective and personal account of a few subjects that I think are interesting or even amazing, which I hope will be enough to reach the goals that I just mentioned. The book uses a wide range of examples from all around the world and from species ranging from vertebrates to bacteria. Nevertheless, I've spent most of the past 45 years studying invertebrates, so it's no surprise that I use a lot of invertebrate examples. Other biologists would use different examples—my friend who has stuffed diatom dolls on her desk would tell you more about freshwater algae; the ichthyologists I know who study fishes all week and then go fishing on the weekends would tell you about their favorite beautiful fish; the scientists I know who wear earrings made by aquatic insects (yes, you read that correctly) would make sure that you heard about the secret lives of caddisflies; and the guy I know who has a coffee cup decorated with pictures of aquatic earthworms would regale you with stories about amazing worms. If any of them read this book, I know that they will scold me for leaving out all of their favorite examples. But no matter what examples we chose, we would all tell the same story, namely, that the wide diversity of inland waters contains a remarkable array of species, some of which are at the edge of extinction or already extinct. If you want to learn more about the good stuff that I've left out of this book, you might want to read some of the references that I cite in the notes for each chapter or the books listed in the further reading section in chapter 15.

THEME AND VARIATIONS

All bodies of inland water are both the same and different from the ocean. They are the same as the ocean because every body of inland water on Earth is made up of water rather than another liquid and because that water contains chemicals that are made up from the same set of elements that the ocean contains. Almost every body of water also contains some form of life. This basic system of water, chemicals, and living things (the ecosystem) is subject to all the usual universal physical and biological laws (conservation of mass, evolution by natural selection, and so on).

Further, all bodies of water on our planet face special constraints because they are on Earth rather than somewhere else, such as Jupiter: they are subject to Earth's gravity (32 feet per second squared [9.8 meters per second squared]) and not Jupiter's (81 feet per second squared [24.7 meters per second squared]), and they have contact with Earth's atmosphere of nitrogen, oxygen, and distinctive traces of other gases rather than to Jupiter's atmosphere of hydrogen, helium, and whiffs of methane, ammonia, and other gases, to name just two examples of differences between the planets. In these very fundamental ways, every body of water on Earth, whether an ocean or a puddle, is alike and can be expected to behave in some of the same ways.

But each body of inland water is different from the ocean and from every other body of inland water. Inland waters vary enormously in their size, shape, age, permanence, geographic location, surroundings, currents and waves, temperature, chemical composition, and vulnerability to floods and other destructive forces. Partly as a result of these differences in physical and chemical conditions, each body of water supports its own collection of life-forms, sometimes including species that evolved in that body of water and are found nowhere else. This individuality in physical, chemical, and biological properties means that each body of water is unlike any other.

The ocean and inland waters differ most obviously in their size and number. The ocean is far larger than Earth's inland waters: it covers almost 100 times as much surface area as all lakes, rivers, and wetlands combined.[4] All of the world's inland surface waters (lakes, including the salty ones like the Caspian Sea, ponds, wetlands, rivers, and streams) together contain less than one-fiftieth of 1% as much water as the ocean. Even if we add the voluminous groundwaters, all inland waters together contain less than 2% as much water as the ocean. No matter how you measure them, inland waters are pretty puny next to the ocean.

But when we *count* water bodies, inland-water bodies rule: there is just a single ocean, but there are perhaps 300 million lakes and ponds larger than a quarter acre (0.1 hectare).[5] On top of this, it is estimated that there are something like 30 to 60 million miles (50 to 100 million kilometers) of rivers and streams. But even these estimates undercount the immense number of inland waters because they leave out wetlands and aquifers (which are hard to count) as well as the innumerable ponds and streams too small to be included in conventional inventories. This multitude of individual water bodies provides a huge range of physical and chemical conditions along with millions of unique sites for evolution to work.

You could think of the ocean as a musical theme with the inland waters as variations on that theme, all played using the same collection of notes and the same suite of orchestral instruments. Although rooted in the theme of an ecosystem in a body of water on the planet Earth, each variation is unique, fresh, and appealing, and capable of revealing something new.

Now let's have a look at some of these variations.

Chapter 2

INLAND WATERS
Types, Sizes, and Shapes

Inland waters are the waters on Earth's continents, large or small, fresh or salty; that is, the blue parts of our planet other than the ocean. It is customary to divide the inland-water world into standing waters (lakes and ponds) and running waters (rivers and streams); sometimes wetlands (marshes, swamps, and bogs) and aquifers (underground formations that hold groundwater) are also recognized as separate categories. Without going into a comprehensive 12-volume catalogue, I want to show you some of the diversity in inland-water habitats that provides such broad opportunities for inland-water life. I begin in this chapter with variation in size and shape, and then in subsequent chapters I describe other kinds of environmental variation that are important to inland-water life.

LAKES AND PONDS

Let's begin our tour with the standing waters: lakes and ponds. Lakes and ponds come in all sizes.[1] Although the ocean is many times larger than any lake, very large lakes like the Caspian Sea, the North American Great Lakes, Lake Baikal, and Lake Victoria have the feel of oceans. These lakes cover more than 10,000 square miles (26,000 square kilometers), so they are too large to see across and can seem to go on forever. Schools of silvery fishes like herrings, whitefishes, and sardines roam their open waters, far from land. Storms on these big lakes produce waves more than 10 feet (3 meters) tall that crash into shorelines and can sink ships—the *Edmund Fitzgerald* is just one of about 10,000 ships wrecked on the North American Great Lakes alone.[2] Indeed, these lakes are large enough to support shipping by oceangoing vessels, and lakeside cities give the impression of the seaside with their working ports, shipyards, commercial fishing, and "seafood" restaurants (since I guess there isn't such a thing as "lakefood"). A few dozen lakes are more than 1,000 feet (300 meters) deep; they give the bottomless impression of the ocean and are home to rarely seen, mysterious animals living in the abyss. Even though these lakes are by any objective measure tiny compared to the ocean, it isn't much of a stretch to describe them as "oceanic."

Very different are the hundreds of millions of ponds of ¼ to 2½ acres (0.1 to 1 hectares).[3] Literally only a stone's throw across and often shallow enough to wade across, they seem like little, toy lakes— no destructive waves, no unbounded vistas, no unknowable depths. Ponds are filled with their own sort of life: duckweed and water lilies, frogs and salamanders, ducks and herons, insects and fairy shrimps, and countless algae, microbes (see digression 2.1), and animals too small for a casual observer to notice.

DIGRESSION 2.1

What Is a Microbe? Imprecise Terminology and Blatant Sizeism

Although I and other aquatic ecologists use the word "microbe" all of the time, it turns out that the term doesn't have a simple, uniformly accepted definition. Microbe means "small life"; often, it is said to be a synonym of "microorganism" and to refer to life-forms too small to be seen without the aid of a microscope. In fact, some small organisms are conventionally excluded from this definition—small animals are almost never included as "microbes," even though some are too small to be seen with the naked eye. On the other hand, fungi are often treated as "microbes," even though the largest living organism on Earth may be a fungus (a single honey fungus in Oregon is thought to cover 2,384 acres [965 hectares]).[4] As far as I can tell, scientists usually use the term to refer to bacteria, archaeans, protozoans, protists, small algae, fungi, and sometimes viruses, although this precise definition isn't universally accepted. In this book, I use "microbe" to mean bacteria, archaeans, protists, protozoans, and fungi.

While you may have heard of bacteria, protists, protozoans, and fungi, many of you probably don't know about archaeans, which weren't recognized as distinct from bacteria until 1977.[5] Archaeans are the size and shape of bacteria, and like bacteria they lack a cell nucleus. But molecular studies show that they are no more closely related to bacteria than we are, and they are now recognized as one of the three major branches of life on Earth (the other two are bacteria and eukaryotes, which includes all organisms that have cell nuclei, like animals, plants, fungi, algae, protozoans, and you and me). Some archaeans can live in extremely hot, acidic, or salty environments, and some can perform functions that no other organisms can, such as produce methane.[6] Scientists don't yet know how many archaean kingdoms or phyla there are—about 20 phyla are recognized at the moment.[7] Despite their small size and low profile outside of scientific circles (I see that my version of MS Word doesn't even know the word, for example), archaeans are a major part of life on Earth and are essential to the functioning of the global ecosystem.

While the term "microbe" might be convenient, it obscures an immense amount of biological diversity: the category includes organisms of very different sizes, evolutionary origins, environmental preferences, and ecological functions. To see how unsatisfactory (and how blatantly sizeist) it is to lump bacteria, archaeans, protists, protozoans, and fungi under the single term "microbe," let me define all organisms larger than 1 millimeter (1/25 of an inch) as "macrobes." I could then accurately write that forests provide good habitat for many macrobes, that macrobes are also abundant in most aquatic environments, and that macrobes play essential roles in both kinds of ecosystems. You might reasonably object that this sentence obscures so much essential detail that it is nearly useless—I've lumped together the owls, oaks, and millipedes of the forest and failed to distinguish them from the mayflies, fishes, and lily pads of the water. What generalizations could I possibly make about macrobes and their roles in ecosystems that would be of any interest? And yet microbes as conventionally defined are far more diverse than my "macrobes."

We lump together small organisms in this way only because we (as large organisms) focus on the big stuff that we can see in our mac-roworld. But we ignore at our peril the enormous differences among different kinds of microbes. Whenever you see the word "microbe" in this book or anywhere else, you should understand what kinds of organisms are included and be more than a little suspicious about what is being lost by the use of such a coarse, sizeist classification.

Then there are the even tinier "lakes" too small to have been counted by scientists and belittled as "puddles" when they are noticed at all: water collected in depressions in fields, floodplains, rough bedrock, tire tracks, tree holes, gutters, and rough sidewalks. Even these tiny and ephemeral bodies of water support ecological functions and their own distinctive forms of life. For instance, puddles in the tire tracks of dirt roads are a characteristic habitat for clam shrimps (technically called Spinicaudata), which are cousins to the more familiar water fleas that we meet in later chapters.[8] These quarter-inch to half-inch (about 1 centimeter long) crustaceans look like swimming clam shells and are rarely found in what you'd think of as a real lake.

The tiniest lakelets are important to human welfare, too. The yellow fever mosquito, the tiger mosquito, the common house mosquito, and other troublesome mosquitoes breed in miniature bodies of water such as puddles and the water standing in discarded tires, old paint cans, and bird baths in your back yard. In addition to being irritating, these "container-breeding" mosquitoes carry terrible diseases such as Zika virus, West Nile virus, dengue, chikungunya, and yellow fever. Campaigns to control these diseases and mosquitoes often begin with removing small artificial lakelets from the neighborhoods where people live. Although they have none of the majesty of the Great Lakes, even these most humble parts of our blue planet support life and are important to us.

Lakes come in all kinds of shapes as well as sizes. Some lakes have steep sides that drop straight into deep water (fig. 2.1). Crater Lake, lying in the caldera of a volcano, is the deepest lake in the United States at an impressive 1,949 feet (594 meters). Even more impressive, Lake Baikal, the deepest lake in the world, reaches a depth of 5,387 feet (1,642 meters) and contains about 20% of all of the liquid fresh water on Earth's surface. Not quite as deep but even more dramatic, Red Lake, which lies in a sinkhole in Croatia, has near-vertical walls that drop almost 800 feet (240 meters) to the lake's surface and another 900 feet (275 meters) beneath the water's surface.[9] There are no expansive beaches or shallow kiddie areas in this lake! Other lakes are so flat that the water may be only waist-deep a mile (1.6 kilometers) offshore. Deep lakes can provide cool refuges in their depths for trouts and other cold-loving species, while the well-lighted sediments

Fig. 2.1. Crater Lake, Oregon, a dramatically steep-sided lake lying in the caldera of a volcano. *Source:* WolfmanSF, CC BY-SA 3.0, Wikimedia Commons.

of shallow lakes can host beds of aquatic plants where many animals find shelter and food.

Some lakes have smooth, regular bottoms, while others are full of islands and hidden underwater reefs and channels. Lake bottoms may be made of bedrock, cobbles, sand, mud, or peat. Likewise, lakes have all kinds of shapes in a bird's-eye view. Some are perfect circles or smooth ovals, inspiring the common, unimaginative "Round Lake," though curiously rarely the name Oval Lake. Others (especially reservoirs) are complicated and spidery, with long arms that reach far into the countryside. Nearly every part of these lakes lies near a shoreline. All of these aspects of lake shape affect what kinds of ecosystems develop in the lake and what kinds of plants and animals can live in it.

RIVERS AND STREAMS

Running waters—rivers and streams—come in a similarly wide range of sizes and shapes. The world's largest river, the Amazon, delivers 20% of all of the water that rivers carry to the ocean. Near its mouth, it is about 30 miles (48 kilometers) wide—too wide to see the other shore. The great rivers of the world offer special habitats for freshwater life: floodplain forests stretching miles across that fill up with fishes during times of high water, strings of shallow backwater lakes that become part of the river during floods, and flowing channels more than 100 feet (30 meters) deep, whose bottoms are darker than any dark you've ever seen.

Some of these rivers have great rapids that must count as among the most awe-inspiring parts of our planet. The Inga Rapids of the lower Congo River, which carry seven times as much water as Niagara Falls, are so powerful that they could supply something like a quarter of Africa's needs for electricity if they were harnessed for hydropower.[10] The Inga Rapids are impassible to boats and thus delayed European colonization of central Africa for centuries. These rapids still have not been well explored by biologists, although modern technology is gradually uncovering their secrets. What the small amount of exploration that has taken place has revealed is that they contain a diverse array of fish and shellfish species found nowhere else in the world that are specialized for life in the torrential rapids. Similar, but smaller, stretches of rapids occur on other great rivers, although many have been destroyed by hydropower dams (the Three Gorges stretch of the Yangtze River, for example) or are planned to be destroyed in the near future (we revisit this topic in chapter 13).

But just as there are many more smaller lakes than larger ones, small streams (creeks, becks, brooks, burns, runs, branches, forks, arroyos, rills and kills, and rivulets) are much more common than great rivers. In the United States, there are more than 1.5 million miles (2.4 million kilometers) of streams categorized as the smallest that are routinely mapped compared to only about 1,800 miles (2,900 kilometers) of the largest rivers, which means that you're far more likely to live near some place like Plum Creek or Floyd's Branch than the Mississippi.[11] And just as with lakes and ponds, many smaller streams have escaped the attention of mappers: tiny spring-fed brooklets and seeps, ephemeral channels that run only part of the year, and human-made "streams" like drainage tiles in farm fields, buried stormwater drains, and even gutters, all of which provide habitat for aquatic life.

Streams and rivers also come in all kinds of shapes. Some are filled with rapids and waterfalls; others are so flat and sluggish that it's hard to know which way the water is going. Some are shallow ("a mile wide and an inch deep"), while others are more than 100 feet (30 meters) deep. Some streams have floodplains that are miles wide, filled with marshes, lakes, wet meadows, wooded swamps, and high-water islands, while others are entrenched in gorges with no floodplain at all. Some flow through cornfields, some through boreal forests or tropical jungles, some through mountain meadows, and others through the downtowns of great cities. Some rivers are full of islands, some are looped into meanders and change direction every half mile (0.8 kilometer), and others are almost as straight as a ruler. Some rivers are fed by many tributaries, while others flow for miles as a single channel. The bottom of streams may be anything from bedrock to cobblestones to clean sand to soupy mud. Taken together, this variation provides an infinite variety of habitats for stream dwellers.

WETLANDS

Now we move from lakes and streams to the black sheep of the inland-water world: wetlands. Most people think fondly of lakes and rivers: we build fancy houses and parks along their shores, spend our leisure hours there, and hang paintings and photographs of lakes and rivers in our museums and homes.

In contrast, our feelings about wetlands have been distinctly less favorable, ranging from distaste to aversion and hostility. The terms we use for these ecosystems hardly evoke warmth or fondness: marsh, swamp, mire, quagmire, bog, fen, morass, often modified by words like oozy, mucky, or slimy. It's easy enough to see how different we feel about wetlands if we do a little word substitution: the Oscar-winning film becomes *On Golden Swamp* and Leonard Cohen's famous lyrics are transformed to "Suzanne takes you down to her place near the quagmire." You get the idea. Lapping *lake* water surrounds Yeats's Innisfree, and a *river* is the center of Huck and Jim's epic journey, but a foul *wetland* is where Pip's convict Magwitch hides. Even in children's books, Viola Swamp is the name of the world's meanest substitute teacher.[12]

These negative attitudes have provoked hostile actions toward wetlands. Although some lakes and rivers have been destroyed as a by-product of our thirst for water for farms and cities (as chapter 13 shows), wetlands are the only inland waters that are commonly destroyed because they are thought to be worthless or undesirable. Wetlands often have been viewed as wastelands—no good for farming, no good for building houses, no good for roads or navigable waterways. So we drain them so they can be farmed, build dams that convert wetlands into reservoirs, or cut channels through them for boat passage.

We've carried out this destruction with enthusiasm and efficiency. Something like 50–90% of the world's inland wetlands have been destroyed since the year 1700, and wetland destruction continues in many parts of the world.[13] In the United States (excluding Alaska and Hawaii), about half of the wetlands have been destroyed, with percentages reaching near 90% in places like California and Ohio.[14] Huge wetlands such as Tulare Lake and other shallow lakes and wetlands in California's Central Valley and the Great Black Swamp in northwestern Ohio have been obliterated so thoroughly that people passing through or even living in what are now extensive agricultural regions have no idea that just a few centuries ago they would have been in the midst of vast wetlands.

For instance, the Great Black Swamp was about 1,500 square miles (3,900 square kilometers, which is about the area of the Florida Everglades that has been preserved) of wooded swamp lying in a broad band southwest from Toledo, Ohio.[15] Its dismalness and impassibility feature prominently in nearly every regional traveler's journal from the early nineteenth century. A road cut through the swamp with great effort in 1827 was said to be "perhaps the worst road on the continent," one that improved transportation only to a limited extent: "hauling stalled teams out of the worst mudholes had become a regular and well-established employment." Finally, the recognition that the soils of the swamp, once drained, were extraordinarily fertile, led to the draining of the swamp between 1850 and 1900. Today, what had been the Great Black Swamp is almost entirely broad fields of corn, soybeans, and wheat. So thoroughly has it been obliterated that I never knew that it existed until I read a book about it, even though I grew up along its edges.

In some cases, obsolete place names provide a clue to the history of former wetland landscapes. Just across the state line from what once was the Great Black Swamp, the hamlet of Ottawa Lake, Michigan, lies amid flat fields of corn, soybeans, and wheat, its shallow, marshy lake long gone (as are the Ottawa, the people who lived in and around the Great Black Swamp before they were forcibly removed to Kansas in the early 1800s; "Ottawa Lake" is thus a doubly anachronistic name). In wet years, the lake reappears, like an arrowhead turned up by spring plowing, to remind us of the land's history (and as I write this in the extraordinarily wet spring of 2023, even Tulare Lake has risen from the dead to flood 30 square miles [78 square kilometers] of reclaimed farm fields).[16]

So if we were classifying the kinds of wetlands, a cynic might start by simply distinguishing between the many ghost wetlands that have been destroyed by humans and the minority of wetlands that have managed to survive to the present day.

Just like lakes and streams, the surviving wetlands are highly var-ied. Here, I focus on size, geographic location, and the kinds of plants that dominate the wetland. Probably the smallest wetland I've ever noticed was next to a perennially leaky fire hydrant in Ithaca, New York: perhaps a square foot (0.1 square meter) of wet ground con-taining a patch of algae and a single wild iris plant, but there must be countless little pocket wetlands small enough to step across. It's hard to estimate the size of the biggest wetland—by their nature, wetlands often have indistinct boundaries, grading imperceptibly into uplands or into open-water ecosystems like lakes, ponds, and rivers. To make the estimation problem even harder, the extent of many wetlands fluc-tuates between seasons and between wet and dry years. Despite these measurement challenges, it is clear that there are some whoppers. The West Siberian Lowlands is a wetland of about a million square miles (2.5 million square kilometers), an area larger than Greenland or Mexico made up of bogs and fens, and the Pantanal, the world's largest tropical wetland, covers about 60,000 square miles (155,000 square kilometers) in Brazil, Bolivia, and Paraguay (table 2.1). In between that leaky fire hydrant and the West Siberian Lowlands, there are countless wetlands of the most varied kinds: prairie potholes, blackwater cypress swamps, cattail marshes along the edges of lakes, papyrus fields along the Nile, reed marshes along river deltas, sedge meadows, and bogs of spruce and *Sphagnum* moss, to name just a few.

DIGRESSION 2.2

What Good Is a Swamp?

The shortcomings of wetlands from the viewpoint of human uses have been obvious for a long time—they're hard to cross on a horse or in a wagon or boat, too wet to grow most of the usual crops, too soggy for building, and often full of disease-carrying mosquitoes. The many benefits that wetlands provide to people are less obvious and have only recently begun to be appreciated and tallied.

Wetlands play key roles in the water cycle. They act as natural flood-control reservoirs, storing water during wet periods of the year, thereby reducing the size of damaging floods downstream. Because floods cause billions of dollars of damage and kill thousands of people per year, the flood-control benefits provided by wetlands (and costs incurred by destroying them) can be very large. The flood waters captured and stored by wetlands are then slowly released, feeding rivers during the dry season or replenishing the groundwater in aquifers, again with large benefits to people and wildlife.

Wetlands can improve water quality by removing nutrients and other pollutants. Indeed, people sometimes use wetlands to treat sewage and stormwater. For example, the city of Gainesville, Florida, was under pressure to reduce the amount of nitrate in wastewater that entered local waterways and ultimately polluted the regional aquifer.[17] Instead of building a brick-and-mortar treatment plant, Gainesville built a complex of wetlands and ponds of more than 125 acres (50 hectares) to treat its sewage effluent and stormwater.[18] This facility, which gives new meaning to the phrase "sewage treatment plants" as marsh grasses rather than industrial facilities, successfully removes nitrate from the sewage effluent, captures sediment and trash from the stormwater, and evens out water flows from stormwater runoff into local waterways. As a bonus, the constructed wetlands are one of the most popular spots around Gainesville for walking and bird-watching. On the January day when I visited, the 40-slot parking lot was nearly full, and dozens of people were out enjoying a visit to what was essentially an attractive sewage treatment plant. Many natural wetlands provide these same nutrient-removal benefits and bird-watching opportunities at no cost.

Of course, wetlands provide a habitat for many plants and animals besides mosquitoes. We harvest some of these plants and animals for our benefit (plants, fishes, waterfowl, crayfishes, and turtles for food, reeds for thatch, and both plants and animals for medical use, for example). A friend who worked in Thailand once described its wetlands as grocery stores for the local people. In addition to using plants and animals that live in wetlands for their whole lives, we harvest many species whose young use wetlands as nurseries and then move into rivers and lakes where we catch them. Although we may think of these species as river or lake fishes because that's where we catch them, without those wetland nurseries there would be fewer of these fishes for us to catch.

Wetlands also provide benefits to people as objects of beauty. Despite their having been regarded as ugly and described as "foul," "dismal," "horrid," and "ghastly" wetlands are now increasingly recognized as beautiful. They have become destinations for hikers, bird-watchers (as in Gainesville), and ecotours, and they are even mentioned in real estate advertisements as local amenities.

The total monetary value of the benefits that wetlands provide is just now beginning to be estimated, but it appears to be very large. One group of experts concluded that "the economic worth of the ecosystem services provided to society by intact, naturally functioning wetlands is frequently much greater than the perceived benefits of converting them to 'more valuable' intensive land use," and another recent study estimated that the inland wetlands of the world were worth more than $3 trillion (!) per year.[19] Whatever their precise monetary value, inland-water wetlands are far from being wastelands.

GROUNDWATERS

If wetlands are the black sheep of inland waters, then our final category, groundwaters, must be the dark matter—large in extent and volume but rarely seen and often ignored. Groundwaters are so obscure that few people (even scientists) would think to include them in a list of inland-water ecosystems. Almost no one has even seen groundwater ecosystems, except for a few spelunkers (who see just a tiny subset of Earth's groundwater habitats when they go into caves). When we think of groundwaters at all, we think of them merely as a source of water for drinking and irrigation, not as ecosystems that support unique species and vital ecological processes. Yet in terms of sheer volume, groundwaters constitute the largest part of the inland-water realm (containing more than 100 times as much water as in the world's lakes, wetlands, and rivers), are vitally important to people, and contain diverse and poorly understood ecosystems.[20]

To appreciate the obscurity of groundwaters, consider table 2.1, which shows the world's ten largest inland-water ecosystems in each major category—lakes, rivers, wetlands, and aquifers. It's likely that you'll know at least some of the world's biggest lakes and rivers (except perhaps for Lake Vostok, which lies under the Antarctic ice sheet, was discovered only in about 1993, and might better be grouped with the groundwater ecosystems than with the lakes).[21] You might know a few of the wetlands, too, by general location if not by the name listed in the table. But only the geekiest of the water geeks will know any of the world's great aquifers (I knew just three of them, and I pride myself on my water geekiness), even though they are far larger than the world's largest lakes. The largest aquifer system on the list, the Russian Platform Basins, covers an area of more than a million square miles (more than 2.5 million square kilometers), which is about the size of India, and is up to 12½ *miles* (20 kilometers) thick. At the other end of the spectrum, small groundwater ecosystems may encompass just a few cubic feet of wet ground.

Aquifers differ from one another in many ways besides size. Many people think of groundwaters as just rivers and lakes that lie

TABLE 2.1

THE WORLD'S LARGEST BODIES OF
INLAND WATER, BY MAJOR CATEGORY

Lakes	*Rivers*	*Wetlands*	*Aquifers*
1. Caspian Sea	1. Amazon	1. West Siberian Lowlands	1. Russian platform basins
2. Lake Baikal	2. Ganges-Brahmaputra	2. Amazon River basin	2. West Siberian basin
3. Lake Tanganyika	3. Congo	3. Hudson Bay Lowlands	3. Amazon River basin
4. Lake Superior	4. Orinoco	4. Congo River basin	4. Atlantic and Gulf Coast aquifer system
5. Lake Malawi	5. Yangtze	5. Mackenzie River basin	5. Lake Chad basin
6. Lake Vostock	6. Rio de la Plata	6. Pantanal	6. Ogaden-Juba basin
7. Lake Michigan	7. Yenisei	7. Mississippi basin bottomland forests	7. Arabian aquifer system
8. Lake Huron	8. Mississippi	8. Lake Chad basin	8. Taoudeni-Tanezrouft basin
9. Lake Victoria	9. Lena	9. Sudd and other Nile River wetlands	9. Nubian aquifer system
10. Great Bear Lake	10. St. Lawrence	10. Prairie potholes	10. Great Artesian basin

Sources: https://en.wikipedia.org/wiki/List_of_lakes_by_volume; https://en.wikipedia.org/wiki/List_of_rivers_by_discharge; Paul A. Keddy, Lauchlan H. Fraser, Ayzik I. Solomeshch, Wolfgang J. Junk, Daniel R. Campbell, Mary T. K. Arroyo, and Cleber J. R. Alho, "Wet and Wonderful: The World's Largest Wetlands Are Conservation Priorities," *BioScience* 59, no. 1 (2009): 39–51; Jac van der Gun, *Large Aquifer Systems around the World* (Guelph, Ontario: Groundwater Project, 2022), doi.org/10.21083/978-1-77470-020-4.

Note: Lakes are ranked by volume, rivers by the amount of flow per year, wetlands by area covered, and aquifers by their area times their maximum thickness.

underground. In areas of limestone and other water-soluble rocks, groundwaters may indeed form large channels or pools that look something like the ponds and streams on Earth's surface. More often, though, groundwaters are held in sand or gravel or flow through cracks and fissures in otherwise solid rocks, so many aquifers are more like an underground sandy beach than ponds and streams. Rates of water flow through aquifers held in such fine-grained materials typically are very slow, sometimes reaching speeds approaching an inch or centimeter per second, similar to a slow-flowing stream, but are more typically in the range of inches or centimeters per day to inches or centimeters per year or even slower.[22] Groundwaters in aquifers where flow rates are so exceedingly slow may be thousands or even millions of years old (that is, water that fell as rain or snow onto Earth's surface thousands to millions of years ago). Drawing water from such aquifers can be more like mining mineral deposits from the ground than pumping water from a rapidly replenished river or lake.

Slow-moving groundwater that is held in close contact with various kinds of rocks and minerals for very long periods of time may, furthermore, develop unusual and distinctive chemical contents based on the peculiar compositions of those rocks, especially if they dissolve readily in water or are radioactive. Because Earth's rocks are so varied, some groundwaters have chemical contents that even lie outside the ranges of the highly variable surface waters discussed in chapter 6.

The distinctive chemistry of groundwaters may provide unusual challenges and opportunities for humans (as well as aquatic life; see digression 11.1). For instance, some groundwaters are brinier than the sea, and others contain too much of naturally occurring chemicals (fluoride, for instance, or arsenic) to be safe for people to drink.[23] On the other hand, the peculiar chemistry of groundwaters may benefit humans. The original Dow Chemical Company, for example, was built on bromide-containing brines pumped from groundwaters beneath Michigan.[24]

Moreover, the tiny pores in fine-grained materials (e.g., silt, fine sand, or unfractured rocks) may be so small that the microbes that live in groundwaters held in such materials cannot move quickly (or at all) through the groundwater (it's like they're living inside a huge filter) and so may have been trapped in place for millions of years.[25] These microbes' home thus is a sort of island, far distant from the rest of Earth, a place where unique forms may evolve, including microbes that may be especially adapted to the unique and challenging physical and chemical environment of their local groundwater ecosystem.

• • • •

From the ankle-deep waters of a puddle in a tire track to the black depths of Lake Baikal and groundwaters more than a mile (1.6 kilometers) below the surface of Earth and from a thin film of water running over a rock face to the miles-wide Amazon River, inland waters offer a nearly infinite range of shapes and sizes to support different kinds of life. But inland waters do not only differ from one another in terms of size and shape. In chapters 4–6, we look at some of the other ways in which inland waters differ from one another (and from oceans). But first let's consider a question that no one ever asks me: where do inland waters come from?

DIGRESSION 2.3

The Mysterious Depths

One of the reasons the ocean is alluring is that it is so deep and vast that it feels like it will never be fully explored. Indeed, each year scientists find amazing new creatures and habitats in the ocean—huge squids, black smokers along the hydrothermal vents—and it seems as if there will always be something fantastic to discover next year.

Do inland waters harbor such deep secrets? With the possible exception of the largest lakes and rivers, it would seem that inland waters must have been pretty thoroughly explored by curious 10-year-olds as well as scientists. Yet as chapters 9–12 show, there are many unsolved mysteries about even the most common of organisms, including those that live in puddles and creeks.

And there *are* vast inland-water habitats that have not yet been well explored by scientists. I'm thinking not about the depths of Lake Baikal (although these depths are indeed mysterious and contain many unique species, as chapter 4 shows) but about groundwaters, which underlie much of Earth's surface. Most people probably think that groundwaters are sterile (unless people have contaminated them). But like surface waters, groundwaters are filled with diverse life-forms. Because groundwaters are so hard to explore, we are still discovering the life that exists in them and will continue make discoveries about groundwater ecosystems for many years to come.

No huge squids, though. Groundwater habitats pose at least two problems for large creatures. First, most groundwater flows through sand, gravel, or tiny cracks in solid rock, so there is no space for big animals to move around. There are a few exceptions, where groundwater flows through caves and large underground channels, chiefly, as already noted, in limestone and other rocks that dissolve easily in water. Here, scientists have found blind, unpigmented crayfishes, salamanders, and fishes (including a little catfish called Satan—biologists think it is cute to name groundwater species after famous underworld figures, so many are named after Charon, Styx, and Cerberus). In addition to the problem of inadequate physical space, most groundwaters are so poor in food (as discussed in digression 11.1) that there is little for large animals with correspondingly large appetites to eat.

Presumably as a result of small pore spaces and scarce food, few animals have been found in groundwaters very far from Earth's surface, but microbes often are found in samples taken miles beneath Earth's surface.[26] Despite this restriction to the shallowest groundwaters, thousands of species of animals have been discovered in groundwaters.[27] In addition to a few fishes and salamanders, the world's shallow groundwaters teem with small crustaceans, worms, and other invertebrates. Many of these are specialized to life in groundwater and do not occur in lakes and streams on Earth's surface, so they are rarely seen by people.

New forms of groundwater organisms are being discovered all the time. I spent a few weeks in the early 1990s looking for groundwater invertebrates in the American Southeast, just scratching the surface with hand tools. This superficial exploration resulted in the discovery of about two dozen new species, including a new order of tiny earthworms.[28]

Microbiologists, however, are making the really exciting discoveries. They have found many new microbes whose remarkable metabolic capabilities are matched to the peculiar circumstances of their groundwater environments. For instance, in 2006, scientists found just a single kind of bacterium living in a sample taken 1.7 miles (2.7 kilometers) beneath Earth's surface, where the water was 140°F (60°C), had a pH of 9.3, and contained no dissolved oxygen or any food from plants on Earth's surface.[29] Despite its grim environment, this bacterium survives nicely by feeding on hydrogen split from the water by radioactive decay in the local rock and can get the carbon and nitrogen it needs from the water. This bacterium is an example of life that does not depend at all on the sun's energy. In the last few decades, scientists have discovered several new phyla of bacteria and archaea in groundwaters, and certainly more remain to be found.[30] Current estimates are that the combined weight of all groundwater microbes is at least ten times as great as the weight of all animals on Earth's surface.[31]

The groundwater realm is so poorly known that we don't even know precisely how big it is (though we know that it is huge—as big as a sea). Nor do we yet know how deep into Earth that life reaches; it has been suggested that living ecosystems may extend as far as 10 miles (16 kilometers) beneath Earth's surface in places.[32] And we certainly haven't discovered all of the life-forms that inhabit this deep and obscure realm, nor how they together contribute to global element cycles. The hidden depths of inland waters are just as mysterious and hard to explore as the deepest ocean abyss.

Chapter 3

ORIGINS
How Inland Waters Are Made

When we're out working on a river or a lake, the people we run into often ask us questions. "Can you eat the fish from here?" Or "how deep is this lake?" Or "how's the river doing?" Or even just "what are you guys doing?" No one has ever asked us where the lake or river came from. I get the impression that people think that lakes and rivers have just always been there, so it doesn't occur to them to ask about where a particular body of water came from. But of course all bodies of water have life stories, some short and simple and others long and intricate, stretching over millions of years. Knowing its origin and life story can tell you a lot about a body of water and the life it contains, so it's worth spending a few minutes considering where lakes and rivers come from.

Let's start with lakes. Lakes and ponds arise when something creates a hole that holds water—that's pretty simple! Their birth can be dated to the time that the hole was created or to the time when the climate changed enough to allow water to accumulate in a formerly dry depression.

Because there are many ways to make a hole in the ground, there are many ways to create a lake—G. Evelyn Hutchinson's famous textbook about lakes lists 96 separate ways that lakes are formed, and I doubt that his list is complete.[1] But most of these causes can be assigned to one of two broad groups: basins made by the blocking of an existing valley and basins made by the creation of new holes.

Dams that block existing valleys are created by lava flows, landslides, ice flows of active glaciers and deposits of sand, gravel, and stone from past glaciers, uplift or faulting of Earth's crust, beavers, and (not the least) people who build structures to make reservoirs for hydroelectricity, recreation, flood control, irrigation water, and so on. These dam-built lakes have shapes and depth contours that are initially set by the shape of the preexisting valley.

Lakes that are made by the damming of existing valleys often have brief lives, for two reasons. First, the valley that is blocked usually contains one or several streams or rivers. Sediments carried by those streams can quickly fill the lake basin, especially if it is small or shallow. Second, the dam holding the lake may be cut away as the outlet stream flows over, under, or through the structure, releasing the water from the lake. This is a problem especially for dams that are made of soft materials (for instance, ice or loose sand, gravel, or rock, haphazardly deposited) or that are not deliberately built and carefully maintained to withstand the pressure and erosive forces of the water.

Sometimes a dam wears down gradually over a period of thousands of years, but it is common for dams to fail suddenly, which can create huge, catastrophic floods. Probably the most famous example of such a catastrophic draining of a dammed lake, at least in the United States, is the Johnstown flood (fig. 3.1). Following heavy rains in 1889, a poorly maintained dam 72 feet (22 meters) tall was abruptly washed away, nearly emptying a lake of 4 billion gallons (15 million cubic meters) in 65 minutes, destroying much of the downstream city of Johnstown, and killing more than 2,000 people.[2]

Fig. 3.1. Johnstown, Pennsylvania, just after the flood destroyed most of the city. *Source:* Steve Nicklas, NOAA Photo Library, Wikimedia Commons.

But floods as large as or even much larger than the Johnstown flood have been caused by the sudden emptying of lakes that were held in place by unstable natural dams. For instance, on about New Year's Day in 1841, an earthquake produced a landslide that completely blocked the upper Indus River in the Himalayas.[3] The Indus being a large river, the valley behind the blockage soon filled, producing a lake 40 miles (64 kilometers) long and as much as 1,000 feet (300 meters) deep. When the water overtopped the dam in June, it failed abruptly. The lake drained in about a day, and "swept everything before it." A witness about 260 miles (420 kilometers) downriver, where the water level rose about 100 feet (30 meters), describes seeing "hundreds of acres of arable land . . . licked up and carried away by the waters" and an army destroyed: "As a woman with a wet towel sweeps away a legion of ants, so the river blotted out the army of the Raja." In chapter

5, we see how even bigger floods resulted from the sudden release of enormous ice-dammed lakes in prehistoric times.

As important as dams are in creating lakes, even more lakes are created by forces that produce new holes in the ground. The most important of these forces are glaciers, volcanoes, the dissolving away of soluble rocks, faulting and other movements of Earth's crust, and people.

The 30% of Earth's surface that was covered by glaciers during the ice age that lasted for more than 2 million years and ended just 12,000 years ago contains many lakes that glaciers made.[4] Only people rival glaciers as lake-builders on our modern Earth, and places like Canada and Scandinavia are dotted with almost countless glacier-made lakes, big and small.[5] It's hard to imagine the size and force of these moving sheets of ice, many of them more than a mile (1.6 kilometers) thick that dragged abrasive stones and boulders along their soles. Glaciers scoured out existing valleys, deepening them to form large lakes such as the North American Great Lakes and the magnificent lakes of the English Lake District and the Italian Alps. Many of these glacial scour lakes are deep and clear.

Kettles are another very common kind of glacial lake. When glaciers melt away, they release huge volumes of sand and gravel, which may contain large chunks of ice from the decaying glacier. If these gargantuan ice cubes, which may be 10 to 100 times as large as a modern sports stadium, are buried by sand and gravel, they may take centuries to melt away, leaving small, steep-sided lakes where the ice cubes had been. Regions where great glaciers melted away, like the Upper Midwest of the United States, are filled with hundreds or thousands of kettle lakes and wetlands.

Volcanoes also make new lake basins. The most spectacular volcanic lakes are the caldera lakes lying in the cones of old volcanoes (Crater Lake in Oregon, shown in fig. 2.1, is a famous example). Caldera lakes form when the emptied lava chamber of a recently active volcano collapses, leaving a large hole. Other volcanic lakes are made when rising lava comes into contact with groundwater, causing large steam explosions. These typically round, deep lakes are common in volcanic districts around the world.

Some kinds of rocks dissolve readily in water. In regions underlain by such soluble rocks (most often, limestone), lakes may be formed where the rocks dissolve away. If large cavities form underground, a sudden collapse may produce a sinkhole containing a lake, probably the closest thing there is to an instant lake. Lakes in regions with soluble rocks are prone to sudden appearance when underground outlets for lake water are blocked and disappearance when those outlets open up. Lake Alachua formed in 1871 in a wet prairie on the outskirts of Gainesville, Florida, when its underground outlet was blocked (allegedly by tourists tossing logs into the outlet to watch them swirl around and then be sucked underground). The resulting lake was large enough to support steamboat traffic, but then vanished abruptly in 1892, dropping 8 feet (2.4 meters) in ten days when the outlet reopened.[6]

Many lakes are formed by movements of Earth's crust. In some cases, uplift or warping of Earth's crust creates shallow basins such as Lake Victoria in Africa or the large, ephemeral lakes of interior Australia. More dramatic are lakes created by the downfaulting of blocks of Earth's crust. Lakes in these downfaulted basins are sometimes bordered by upfaulted blocks and include some of the largest, deepest and oldest lakes in the world, such as Baikal in Siberia, Tanganyika in Africa, and Tahoe in the western United States. Evolution in these very old lakes has produced many unique species, so they have been exceptionally important for the biodiversity of inland waters.

Finally, humans have excavated many new lakes, as well as producing lakes by damming valleys. In recent times, humans have come to rival glaciers as the most important creators of lakes.[7] Human-excavated basins are usually small, with simple geometric shapes and uncomplicated shorelines. Common examples include farm ponds and urban stormwater retention ponds.

Because anything that can make a hole can potentially make a lake, there are many other ways to make a lake, ranging from the mundane (wind erosion that produces broad, shallow lakes, the hoof prints of livestock that generate tiny lakelets) to the exotic (meteorites that form craters or erosion that produces the plunge pools of extinct waterfalls resulting in deep, round lakes).[8] There have been lakes on Earth since the time billions of years ago when water condensed out of the cooling atmosphere, and there will be lakes as long as there is liquid water on the surface of the planet. However, because glaciers (present over the last 2 million years, especially in the Northern Hemisphere), people (active especially over the last 100 years), and beavers (present over the last 10–15 million years, almost entirely in the Northern Hemisphere) have recently been making so many lakes, we are now in an Age of Lakes (the "Lacustricene"?), especially in the Northern Hemisphere, where there must be more lakes than throughout most of Earth's history.[9]

In contrast to lakes, rivers have life stories rather than origin stories. Rivers begin to form when rain (or melted snow) starts to run off of land newly emerged from the sea (or from under a glacier) and evolve along with their landscapes. It doesn't require a special event to make a river, just too much rain or snow to evaporate or soak into the ground. Except for the driest deserts, most of Earth is wet enough to produce runoff, at least occasionally, so streams and rivers that run at least part of the time are nearly ubiquitous on our planet.

What's more, running water has been a major force shaping the surface of Earth. A geologist visiting from another planet would be able tell at a glance that Earth is a planet of rivers. Although people may think that rivers run through valleys that they find, in many cases it is more accurate to say that they run through valleys that they shape.

Of course, forces other than rivers often shape those valleys and their rivers as well, so rivers don't always form those perfect tree-shaped patterns (fig. 3.2), with smooth courses from headwaters to mouth. Here are just three examples. First, here in Michigan and in other areas that were covered by glaciers, thick ice sheets scoured out valleys and left enormous piles of debris that produced an irregular land surface. In the few thousands of years since the glaciers left, the rivers haven't had the time to build regular, tree-shaped drainages. Consequently, rivers in these regions often follow quirky, tortuous paths, and waterfalls are common. Second, in places where hard rock at the land's surface is underlain by softer rock, a river often will produce a waterfall or rapids at the point where the river cuts its way through the hard rock at the surface to reach the softer rock beneath. The waterfall is produced when the soft rock downstream erodes more quickly than the hard rock upriver. Niagara Falls is a dramatic, well-known example. Third, in areas underlain by soluble rocks such as limestone, entire rivers may disappear down holes (evocatively called swallow holes or swallets), into underground channels, only to emerge miles away as large springs.

It is true that rivers in old landscapes running over very uniform geological structure may have built the valleys that they run through and form perfect, tree-shaped drainage patterns with very regular courses. However, many rivers encounter geological events (glaciers, earthquakes) or structures (outcroppings of hard rocks, holes in soluble rocks) that affect the shapes of their valleys, their courses, and the kinds of habitats that they provide to riverine life.

Fig. 3.2. Tree-shaped river drainage patterns in Tibet, as viewed from space. The white is snow cover, which has melted in the valleys. *Source:* NASA/JPL-Caltech.

In the case of rivers, we have very ordinary processes (differential erosion, dissolution of rocks) acting over long periods of time to produce extraordinary features like waterfalls and disappearing rivers. In many lakes, in contrast, dramatic forces (volcanoes, enormous steam explosions, earthquakes, miles-thick sheets of moving ice, ice cubes the size of football stadiums, and even meteorites) produce features that we regard as ordinary (lakes and ponds).

Wetlands may be produced if the same forces that produce lakes or rivers result in bodies of water that are shallow, well-lighted, and slow-moving enough to support vascular plants. As with lakes and rivers, different forces produce different kinds of wetlands: vast marshes that fill the margins or the entire extent of shallow basins produced by the damming of flat valleys, crustal warping, or glacial scour; complicated mosaics of wetlands in the oxbows, floodplain lakes, and margins of great rivers; prairie potholes in glacial kettles; little beaver ponds.

The geologic formations that hold groundwaters have life stories too, which affect their characteristics and the life that they contain. Of course, the chemistry of the groundwater depends on the chemistry of the geologic formation that holds it—groundwaters in limestone, a rock made of calcium and carbonate, are rich in calcium and bicarbonate because limestone dissolves readily, whereas groundwaters in sandstone tend not to contain many dissolved chemicals because sandstone is made of silica and is not very soluble. But the origin and history of the formation also determines how much water it holds and how freely that water can move, both factors of vital importance for groundwater organisms and ecosystems. For instance, mud newly settled at the bottom of the ocean often contains 25–70% water, as does sand freshly deposited by water or wind.[10] Because the water-containing pores are so much larger in sand than in mud, though, water can move thousands of times faster through sand than through mud. After time, heat, and pressure turn these muds and sands into shale and sandstone, respectively, their water contents drop significantly (often to less than 20% and in certain cases less than 5%), and water movement is greatly reduced. Imperfections in the rock, as well as the compactness of the rock itself, affect groundwater. If the rock is fractured, groundwater may move freely along the cracks. Because limestone dissolves in water, groundwaters moving through limestone can produce large caves and passages, large enough to allow for animals such as small crustaceans and even fishes to live and evolve in. Depending on their origins and histories, geologic formations that hold groundwater may contain anywhere from less than 1% water to more than 50% water. These formations may contain large passages through which water moves freely in underground rivers and can support fishes or be so compact that even bacteria cannot move through the rocks. So just as for lakes, knowing the origin and history of groundwaters can tell you a lot about what kinds of organisms are likely to live there and how the ecosystem might work.

The forces that produce bodies of inland waters vary regionally—Canada and Norway were overrun by glaciers, Florida and Slovenia are underlain by soluble limestone, the Rift Valley of East Africa is an active fault zone, there are active volcanoes in Iceland and extinct ones in western Germany, and beavers live in North America and Europe. As a result, the world's lakes are arranged in districts where multiple lakes of a similar kind tend to occur together—glacial scour lakes and kettles in Canada and Norway, sinkholes in Florida and Slovenia, deep rift lakes in East Africa, calderas, lava-dammed reservoirs, or explosion craters in Iceland and Germany, and innumerable beaver ponds in North America (at least before European trappers arrived). And, of course, farm ponds and reservoirs wherever people live.

Because lakes of similar origins tend to have similar shapes and sizes, the lakes in a district often tend to resemble one another in size, shape, age, and habitat types, and so they often support similar kinds of organisms. The deep, glacial scour lakes of Canada and Norway support trouts and other cold-water fishes, whereas the smaller and shallower kettle lakes of glaciated regions offer habitats for lily pads and sunfishes. The large ancient lakes of the East African Rift Valley contain hundreds of unique species that evolved in them, while the beaver ponds of North America and Europe, though teeming with life, support no such unique species.

Knowing the origin and life story of a body of inland water, you can make some pretty good guesses about the size, shape, age, inhabitants, and ecological functioning of that ecosystem. If you ever find yourself on a game show called *What's My Lake?*, you could do a lot worse than beginning your questioning with "where did this lake (or river) come from?"[11]

Chapter 4

AGE

Lifespans of Inland Waters

We've seen that inland waters come in a wide range of sizes and shapes. But differences in size and shape are not the only factors that contribute to the high diversity of inland-water habitats. In the next few chapters, we consider a few other factors, concentrating on things that matter most to the species that live in inland waters.

Let's begin with age. Once we start thinking about bodies of inland water as having origins, it's natural to wonder *when* they were formed, how old they are. The age of a body of water is important to the kinds of species that live there. The most ephemeral of inland waters—say, a road puddle that is filled by the highway department after a week—exist for such a short time that few species can colonize them before they disappear. Many ordinary ponds and lakes last for a few thousand years, long enough for many species to find and colonize them but usually too short a time for evolution to produce many new species of plants or animals in that body of water. Finally, the oldest lakes and rivers may last for millions of years, long enough for evolution to produce many species that are uniquely adapted to the conditions in that body of water.

All lakes have brief lives in terms of geological time. Most lakes last no longer than a few thousand years, whether they were created by glaciers, geological faulting, landslides, the dissolving of underlying bedrock, beaver dams, or, nowadays, humans. They disappear when they are filled with sediment, when their outlet stream erodes itself down into the land and drains the lake, or when the dam holding the lake fails. The thousands of lakes in places like Minnesota and northern Europe were created during the last ice age less than 20,000 years ago, and the many reservoirs and farm ponds that dot modern landscapes are far younger. In a short time (geologically speaking), all of these lakes will disappear. Pretty soon, Minnesota will have to face up to this ugly reality, and change its slogan every few thousand years or so (I'm not sure which would be least embarrassing: "Land of Nine Thousand Lakes," "The State Formerly Known as the Land of Ten Thousand Lakes," or maybe "Land of Ten Thousand Lakes and Soggy Depressions").[1] At the short end of the spectrum, lakes formed by beaver dams or landslides across large rivers may last for a few years or less, and farm ponds and reservoirs built on sediment-laden rivers may last for a few decades. (Although few people would call them "lakes," small puddles are even shorter lived, lasting for only hours to days.) At the other end of the spectrum, lakes built and maintained by ongoing geological faulting may last for millions of years. Lake Baikal in Siberia is 25–30 million years old.[2] However, only about 20 lakes in all of the world are thought to be more than a million years old, and these ages are the blink of an eye compared to the 3.8 billion-year age of the ocean.[3] All lakes are very short-lived compared to the ocean.

DIGRESSION 4.1

What Is the Oldest River in the World?

It's usually possible to know the age of a lake by using radioisotopes or other tools to date the sediments that have accumulated in its bottom. Most scientists agree that Lake Baikal is the oldest lake in the world (there is some evidence that Lake Zaysan in Kazakhstan may be even older, but Baikal holds the title for now).[4] It's much harder to say how old a river is or to decide which river in the world is oldest.

To answer this question, we first have to distinguish between two uses of "age": the age of a section of river channel and the age of a flowing river system. Geologists sometimes are interested in the age of a bit of river channel, that is, how long a particular section of channel has been used. Using this definition of age, the oldest river in the world is often said to be the Finke, an intermittent river in interior Australia, parts of which are thought to be 350–400 million years old.[5] Several other rivers may be more than 100 million years old (and ironically, the New River is supposed to be the oldest river in North America, at perhaps 300 million years).[6]

This view of river age is not so interesting to an ecologist, though. The Finke and other ancient rivers may have carried water for only part of their lives (indeed, the Finke is dry most of the time in today's climate). If a river channel goes dry for, say, a few million years (or even a month), that may not trouble a geologist, but it certainly will trouble a fish. So an ecologist (and a fish) are more interested in how long has a river been flowing (or at least contained water) than whether the channel has remained in one place. As long as there is sufficient precipitation and runoff, water will be flowing through a channel and providing habitat for aquatic organisms. That channel may move around as the river changes its course, or may wax or wane as climate change and tectonic change adjust the amount of water in the channel. I have not seen lists of the oldest rivers according to this ecological definition of age, but many river systems have contained water for millions of years, long enough to allow evolution to produce new species and much longer than most lakes.

Ancient lakes often contain species that evolved there and are found nowhere else in the world. Scientists have found about 1,000 unique species in Lake Baikal that are thought to have evolved in the lake, including the world's only fully freshwater seal, giant amphipods (most freshwater amphipods are less than ¼ inch [6 millimeters] long, but Baikal's bruisers are a couple of inches [5 centimeters] long), and freshwater sponges several feet tall.[7] Surely more of these unique species remain to be discovered in Baikal. Similar flocks of species evolved in other ancient lakes such as Tanganyika in Africa, which is about 10 million years old and 12,600 square miles (33,000 square kilometers) and contains around 600 unique species, Titicaca in South America, which is 3 million years old, 3,300 square miles (8,500 square kilometers), and home to about 70 unique species, and Biwa in Japan, which is 4 million years old and 260 square miles (670 square kilometers) and boasts about 50 unique species.[8] In contrast, the North American Great Lakes in North America, which cover 94,250 square miles (244,000 square kilometers) but are less than 15,000 years old and which together hold about the same volume of water as Baikal, contain fewer than 10 known unique animal species thought to have evolved in the lakes.[9]

The size of a lake is usually roughly correlated with its age and permanence. Lake Baikal is both the oldest and largest freshwater lake in the world, and most of the 20 lakes that are more than a million years old are also among the 1,000 largest lakes. However, there are many exceptions to this pattern. As just mentioned, the North American Great Lakes are about as big as Baikal (in terms of the amount of water that they hold) but just a few thousand years old. Kati Thanda–Lake Eyre in Australia covers more than 3,600 square miles (9,300 square kilometers) when it is full, making it the nineteenth-largest lake in the world. It lies in the arid interior of Australia, though, so it is dry most of the time. Despite its status as one of the world's largest lakes, each incarnation of Kati Thanda–Lake Eyre lasts only for a year or two.[10]

The situation in running waters is very different from that of essentially ephemeral lakes. Rivers are born when rain or snowmelt starts running off of land newly emerged from the sea, and they continue to exist as long as precipitation in the watershed provides enough water to keep the stream running or until the land is submerged by glacial ice or the sea. The rivers in parts of North America, Europe, and Asia that were covered by ice age glaciers are as young as the North American Great Lakes and generally lack unique species. In other parts of the planet, though, many rivers are millions of years old and contain numerous unique species that are found nowhere else in the world. Thus, the Mekong, the Amazon, the Mobile, the Tennessee, the Nile, the Parana, the Colorado, the Congo, and dozens of other old river systems each have dozens to hundreds of unique species that especially adapted to conditions in that river system.

The existence of unique species in most ancient bodies of inland water complicates conservation efforts. If we want to protect the entirety of inland-water life, it is not sufficient to protect a convenient 30% of the world's inland waters, choosing waters that are far from damaging human activities, which are easy to protect. Instead, we must think about protecting each of the ancient bodies of water that contains its own group of unique species, even if some of these ecosystems are right in the middle of dense human populations and difficult to protect.

• • • •

Inland waters range in age from the most fleeting of waters, whose age is measured in days or even hours, to rivers that are hundreds of millions of years old. Young waters and old waters support different kinds of species, and each separate ancient body of water, whether a lake or a river system, typically contains many species that evolved there and are found nowhere else on Earth. But whether young or old, we learn in the next chapter that not all inland waters lead serene lives.

A Brave New World

Before we leave the subject of age, it is worth noting that human activities are short-circuiting the importance of age in determining what species live in a body of water. First, humans move a lot of species from one body of water to another. We deliberately move species that we think are desirable (for example, food fishes like common carp, sport fishes like largemouth bass and rainbow trout, biological controls like grass carp and black carp, furbearers like muskrats and beavers, attractive species like purple loosestrife, water hyacinth, and lotus), we dump unwanted pets, bait, or aquaria, or we are careless and move hitchhikers around on boats and other equipment. I discuss the harmful consequences of these actions in chapter 13 when I consider biological invasions. Regardless of their ecological and economic effects, our actions move species into inland waters that they would have reached on their own only after millennia (for example, the smallmouth bass native to the Mississippi River and St. Lawrence River basins now in the lakes of Maine) or not at all (smallmouth bass now in South Africa). Over time, inland waters will increasingly contain species that someone thought was useful or cute, along with species that are good at hitchhiking with people, rather than species that were able to disperse over the lifetime of a body of water.

Furthermore, although you may not think of people as driving the evolution of inland-water species, there are a few examples of this from the past, and this may become a big deal in the near future. People have bred distinctive varieties of goldfish for centuries, for example, some of which have been released into inland waters, where they have established populations. You can see bright orange and jet-black goldfish swimming around in lakes and rivers. Further, it is likely that long-running, intensive fisheries affected the evolution of fish species.[11]

But we are on the cusp of a revolution in synthetic biology (a field that uses technology to change the functions of existing organisms or build new organisms from scratch), which will result in the deliberate or accidental release of novel organisms into inland waters. This technology is expected to become widely available (by "widely available," I mean like to high school students and enthusiastic amateurs).[12] It is not clear how the production, containment, and release of genetically engineered organisms will be regulated. Based on our experience with biological invasions, you might guess that regulation will be poor and patchy, resulting in the release of new kinds of organisms into inland waters. These new organisms could be something like carp that glow in the dark, but could also be species capable of fundamentally transforming the ecosystem, with wide-reaching ecological and economic effects (we're about to find out, aren't we?). Again, the whims of the humans who engineer new life-forms and the idiosyncrasies of regulations may come to override considerations of lake or river age and species dispersal in determining which species live in a body of water.

Chapter 5

DISRUPTION
Stability and Disturbance in Inland Waters

Although inland-water ecosystems may exist for millions of years, they can be turned upside-down by severe disturbances such as drying or flooding that kill many of their inhabitants. The frequency and severity of such disturbances vary greatly from one ecosystem to the next. Many waters are very stable—think of lakes that never dry up and maintain nearly constant water levels. Likewise, springs and spring-fed streams may not experience dramatic floods or periods of drying, with the result that water levels remain very steady.

On the other hand, most streams and rivers are subject to floods that raise water levels, increase current speeds, tear up stream bottoms, and kill the stream dwellers or wash them downriver. These can be large disturbances—the local river where I go fishing rises more than 8 feet (2.4 meters) during floods, changing abruptly from a shallow river that I can wade across to a muddy torrent that uproots full-grown trees and sends them hurtling down to the lake. The largest known river floods unleashed almost unbelievably enormous forces on these ecosystems (see digression 5.1).

Conversely, even large streams can stop flowing altogether during dry times, shrinking to isolated pools or dusty river beds. During such times, fishes and other aquatic plants and animals will die if they cannot reach a refuge, such as the damp sediments beneath a dry riverbed or a stagnant pool or if they don't have a drought-resistant life stage (chapter 10 describes some of these resistant life stages). Even these refuges may not offer reliable protection; animals living in dwindling pools may be exposed to deadly peril from predators like wading birds and raccoons.

Periods of flooding and drying can be very predictable, as in climates where river flows are driven by annual monsoons or snowmelt. But in other places, such periods are unpredictable; they occur at irregular intervals and can happen at any time of the year (fig. 5.1).

And although you may think of lakes as being more stable than these fickle rivers, lakes and ponds dry up, too. In wet climates like eastern North America and Europe, lakes seem like permanent features. But don't forget that smaller ponds dry up during the summer even in these climates. We even have a special name for these—"vernal pools." They contain water only during the spring and support distinctive animals like fairy shrimps, frogs, and salamanders. And in climates where rainfall is highly seasonal or unpredictable, even very large lakes dry up or shrink to small pools during dry periods. The most spectacular examples probably are from interior Australia, which contains huge ephemeral lakes that are dry salt flats most of the time, and fill to their full extent only rarely. Kati Thanda–Lake Eyre, one and a half times the size of Delaware, filled three times in the last century, and Lake Torrens, which covers 2,300 square miles (5,900 square kilometers), filled just once in the last century.[1]

Because wetlands are so shallow, they may completely disappear and reappear with even small changes in water level from season to season or from one year to the next. There are many cases of these Houdini-esque wetlands—wetlands lying in the floodplains of large

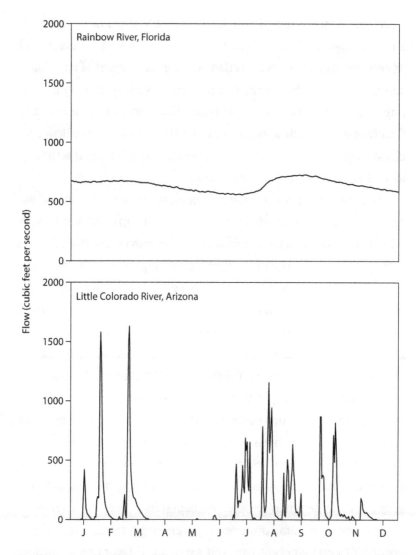

Fig. 5.1. Two rivers with very different patterns of flow. The Rainbow River in Florida is fed by large springs, and its flow is very stable, with almost the same amount of water flowing down the river every day of the year. Organisms in this river don't have to worry about floods or droughts. The Little Colorado River in Arizona is a desert river fed by irregular rainstorms and subject to flash floods. Organisms living here have to deal with long periods without flow punctuated by occasional violent floods. *Source:* Graph for both sites based on 2015 data from the US Geological Survey.

rivers and in shallow basins in highly variable climates are good examples. Other kinds of wetlands have nearly steady water levels, such as groundwater-fed bogs or wetlands along the margins of large lakes, and rarely if ever disappear. The pattern of wetting and drying is so important to the character of wetlands that human actions that artificially stabilize or destabilize water levels in them, even a little, can cause large changes that affect what kinds of species can live in them as well as their ecological functioning.

The high incidence of habitat instability distinguishes inland waters from the ocean, which does not flood or dry up capriciously. Variation in the degree of habitat instability also provides both challenges and opportunities to inland water life, favoring different kinds of lifeforms in different kinds of habitats. In very stable habitats, there can be an advantage to holding on to choice habitats and passing them along to your children. An environment like this fosters longevity, a strong competitive ability, and large, well-developed offspring that are able to fend for themselves. Think of the CEO of an established company grooming his daughter to take a robust family business into the fourth generation. When the habitat is frequently disturbed, however, there is no family business to pass down—the family homestead, the factory, and the head offices might be swept away next week. Instead of concentrating on holding on to a spot, an organism living in an unstable environment may be better off being able to move to a new place when its present home disappears, developing a resting stage (like a cocoon or cyst) where it can wait safely until favorable conditions reappear, or having many offspring, in the hopes that some of them will find a good, unoccupied habitat. The range of disturbance regimes in inland waters, from very stable to highly ephemeral, encourages the evolution of a wide range of kinds of species, each of which can cope with a particular kind of such regimes.

J Harlen Bretz and the World's Biggest Floods

I can't leave the subject of floods without telling you about J Harlen Bretz and the gigantic floods that he discovered.[2] Bretz was a geologist who grew up in Michigan, then moved to the Pacific Northwest in 1907. He quickly became fascinated by the region's landscapes, especially the "scablands" of the Columbia Plateau, arid landscapes dramatically scarred with rugged channels and cliffs. After hiking through the scablands and studying topographic maps, Bretz decided that all the evidence pointed to huge floods as the sculptor. The channels are shaped like immense rivers and contain features (fig. 5.2) that look like dry waterfalls, plunge pools, and immense "gravel bars" composed of well-rounded basalt boulders up to 10 feet (3 meters) in diameter. Bretz published his flood hypothesis in 1923, to a swift and savage reception from the geological community. Where did Bretz imagine that all this water came from in the dry landscape of eastern Washington? No such source was anywhere in sight. (Geologists were also understandably a little sensitive on the subject of big floods and may have worried that someone would next announce that they'd discovered an ark just outside of Spokane.) Instead, geologists much preferred the more conservative idea that the scablands were produced by ordinary erosion working slowly over a long span of time.

But Bretz stuck to his guns, and evidence supporting his hypothesis built up over the next few decades. Studies by Joseph Pardee showed that a lake (Lake Missoula) the size of modern Lake Ontario had formed in western Montana toward the end of the last ice age that was held in place by a 2,000-foot-thick (610-meter-thick) ice dam. This dam failed abruptly, sending all that water rushing across the area. Aerial and satellite photos clearly show giant ripples and other flood features. By the

Fig. 5.2. One of the channels cut through the scablands of the Columbia Plateau, Washington State. *Source:* Steven Pavlov, Wikimedia Commons.

1970s, the flood origin of the scablands was finally accepted by the geological community.

Bretz lived to be an old man, old enough to be vindicated. When he received the prestigious Penrose Medal from the Geological Society of America in 1979 at age 96, he said "All my enemies are dead, so I have no one to gloat over."

Geologists now believe that Lake Missoula filled and drained many times, producing perhaps 40–80 floods that tore across the region. The largest of these floods was 10–80 times as large as the Amazon River, carrying perhaps 10 times as much water as the combined volume of all of the rivers in the world today. Current speeds would have reached 80 miles per hour (130 kilometers per hour). The largest floods released up to 1.9×10^{19} joules of potential energy (this is equal to 4,500 megatons of TNT, or 1,500 times the combined force of all explosives used in World

War II). In total, the floods eroded 50 cubic miles (210 cubic kilometers) of material from the area that is now eastern Washington.

But the most startling thing that I discovered in reading about the Missoula floods was the passing comment that they were the second-largest known floods. Wait a minute—the *second* largest? It turns out that there were even larger floods, also resulting from the sudden failure of ice dams that held back large lakes, in the rivers of the Altai Mountains near the border of Russia, China, Mongolia, and Kazakhstan in central Asia.[3] The landscape features left by these floods give us some idea of their immense power. The modest rivers that run through the modern remnants of the flood channels today are bordered by towering terraces of stone and gravel built by raging flood waters. Plains that once were the bottoms of ice-dammed lakes are decorated with ridges that look just like the ripples that you've seen in the shallow waters off sandy lake beaches, only they're *60 feet (18 meters) tall*.

It's hard to imagine that many plants and animals living in the path of such floods could have survived, so these rivers must have been recolonized by organisms living in refuges in unaffected tributaries.

_____

Chapter 6

MATERIALS
The Chemical Diversity of Inland Waters

I know people who really like water chemistry. They spend all day thinking about redox reactions and mass balance and valences and solubility indices and spiraling metrics, and when the workday is over, they go out for a beer with their friends and talk about redox reactions and spiraling metrics. (In my experience, water chemistry enthusiasts are often beer connoisseurs as well, which makes sense in a weird way if you think of a glass of beer as a special kind of aqueous solution.) These are people who when asked to name their favorite chemical element say "ooh, ooh, can I have three?" and then name five. I'm guessing that you're not one of those people.

So instead of going into great detail, element by tedious element, about the enormous variation in the chemical content of inland waters, I'm just going to briefly talk about how much pH varies across inland waters, assume that is sufficient to make my point about the chemical diversity of inland waters, and move on to subjects that you like better than water chemistry.

You may remember from high school chemistry that pH is a measure of whether a substance is acidic or basic (or "alkaline"). Materials that are neutral (neither acidic nor basic) have a pH of 7, acidic materials have a pH less than 7 (household vinegar has a pH of about 2.5), and basic materials have a pH greater than 7 (household ammonia has a pH of about 11.5). The pH scale is logarithmic—a change in pH of one unit represents a 10-fold change in acidity (technically, a 10-fold change in the *activity* of hydrogen ions). So vinegar at a pH of 2.5 has about a billion times more hydrogen ion activity than does ammonia at a pH of 11.5.

The logarithmic scale allows us to conveniently express enormous differences in chemistry but makes it easy to forget that small differences on the pH scale can mean big differences in chemistry which can have large consequences. For instance, the ocean today has a pH of around 8.1, which tells us that it is a little basic. Higher concentrations of carbon dioxide in the air resulting from fossil fuel burning have caused the ocean's pH to fall 0.1 units from a preindustrial value of 8.2, and models suggest that it may fall to 7.8 by the year 2100.[1] These sound like small changes hardly worth worrying about. But a change from 8.2 to 8.1 represents an increase in hydrogen ion of 26%, and a change from 8.2 to 7.8 means an increase of 150%. These changes are enough to cause serious problems for marine life. Organisms like clams and corals that make their shells out of calcium carbonate find it increasingly difficult to build and maintain their shells if the pH drops just a few tenths of a point. Ocean scientists are now scrambling to understand and find ways to prevent or manage these changes before we lose important parts of ocean ecosystems as the pH drops by 0.3 or 0.4 units.

The pH range across inland waters is far greater than the few tenths of a point that are so important in the ocean. This exposes inland water organisms (and chemical processes) to an enormous range of chemical conditions. Most inland waters have a pH between 4 and 9.

Again, this range may sound modest, but it represents a 100,000-fold range in hydrogen ion activity. And there are waters that lie outside even this expansive range.

The most acidic natural inland waters are the lakes that lie in the craters of volcanoes, like Kawah Ijen in Indonesia (fig. 6.1). These lakes are so rich in sulfuric acid that they may have a pH as low as 0.1.[2] To put this in context, fresh battery acid has a pH of about 0.7. The label on battery acid (which remember is about a quarter as strong as this lake water) warns that it causes severe skin burns and eye damage and advises consumers to use personal protective equipment, to immediately call a poison control center if it gets swallowed, and to dispose of it in an approved waste disposal plant. You might feel pretty confident in guessing that nothing lives in this lake.

But when scientists sampled Kawah Ijen (an undertaking that required special gear; as you might imagine, a lot of regular gear like aluminum boats would dissolve in the lake water), they found a green alga and three kinds of archaeans living in the lake.[3] Apparently, no animals live in the lake. However, the acid water in its outlet stream is gradually neutralized as it flows downstream, and the researchers discovered fly larvae called chironomids living in the stream at the point where the outlet stream reached a pH of about 2.5 (like vinegar, remember?).

And more remarkably, not only do these species survive in highly acidic waters but some of them even prefer these harsh conditions. One of the archaeans living in volcanic waters can tolerate pH *below* 0 and grows best at a pH of 0.7.[4] That is, battery acid is its ideal pH and vinegar and lemon juice are far too mild for its taste.

(If you're wondering about how low the pH of nonnatural waters can go, a pH as low as -3.6, yes that's *minus* 3.6, has been recorded in some groundwaters in California polluted by mining wastes.[5] It was

Fig. 6.1. Lake Kawah Ijen, which lies inside a volcanic crater in Indonesia. This lake has a pH of 0.3, more acid than battery acid, but it nevertheless supports life. *Source:* CEphoto, Uwe Aranas.

a major technical problem for scientists to figure out even how to measure such low pH.)

At the other end of the spectrum, alkali lakes often have a pH of 9.5 to 11.5. Alkali lakes typically occur in regions that are so dry that any water that runs into the lake leaves by evaporation rather than through an outlet stream. This allows minerals dissolved in the water to build up to very high concentrations. Depending on the surrounding geology, such lakes may develop into salt lakes (like the Great Salt Lake in Utah) that are filled with sodium chloride (ordinary table salt) or alkali lakes that contain a lot of sodium carbonate (washing soda) and other minerals that give the lakes such high pH and alkalinity. Alkali lakes and the salt flats that form when the lake dries up altogether have been in the news lately because some alkali lakes and flats are a major source of lithium, which is needed to make batteries for electric cars and which has other uses. Alkali lakes also show up in the old westerns—parched travelers who ran out of water two days ago come across a desert pool, and the greenhorn in the group throws himself into the brackish water, swallowing it in huge gulps. Then he stumbles away from the water, retching, after which the laconic leader of the band remarks: "bad water."

Again, you might think that such bad water wouldn't support life. As is the case of the corrosively acidic volcanic lakes, few species other than microbes can tolerate the harsh conditions in alkali lakes, but these few species can be enormously productive. In fact, alkali lakes can be among the most productive of inland waters in terms of the sheer amount of biomass that is grown each year.[6]

So in contrast to the pH of the ocean, which is very near 8.1, the pH of inland waters spans a range from about 0.1 to 11.5, representing a 250 billion–fold range in hydrogen ion activity. Life exists and even thrives over this enormous range. This huge range in pH presents both a challenge and an opportunity for evolution to produce species whose ecology and physiology are adapted to some specific part of it: different species for strongly acid waters, mildly acid waters, neutral waters, mildly alkaline waters, and strongly alkaline waters.

DIGRESSION 6.1

Tea, Anyone?

The chemistry of inland waters is extraordinarily variable, encompassing a range of literally a billion-fold in concentrations of some chemicals from one body of water to another. *Why* is chemistry so variable across inland waters? To understand this variability, it may be helpful to think of the watershed as a sort of a huge tea bag that produces the cup of tea that is the downstream lake or river.

What you get in a cup of tea depends first on what's in the tea bag—orange pekoe, Earl Grey, or chamomile. Likewise, what you get in a body of water depends on what's in the watershed. If the watershed is underlain by materials like hard rocks or quartz sand that don't dissolve readily in water, then the water in lakes and streams will be relatively pure, containing low concentrations of dissolved minerals ("hard rocks make soft water"), while if the watershed contains a lot of limestone, gypsum, rock salt, and other materials that do dissolve readily in water, the lakes and streams will be rich in dissolved materials such as calcium and sodium. And as I've just noted, volcanoes (or metal smelting plants) in the watershed can release a lot of sulfuric acid, producing impressively low pHs in nearby lakes and streams.

Land use and vegetation in the watershed matter, too. If the watershed is filled with animal feedlots and factories, the downstream lakes and rivers are likely to be rich in nutrients and other pollution. If the watershed contains many wetlands, the lake or river may look a lot like an actual cup of tea (but not taste like one!), stained dark brown with dissolved organic molecules leached from the decaying wetland plants. So the most important reason that water chemistry differs across different inland waters is that different things are in their tea bags.

But contact time matters, too. If you leave a tea bag in the cup for an hour instead of two minutes, the tea will become dark brown and undrinkable. Likewise, a watershed with deep soils and aquifers, where the water dwells for decades to millennia before reaching lakes and rivers, will produce waters that are much stronger than a watershed where the rain runs off of bare bedrock in a matter of minutes to hours, even if the two watersheds are made of the same kinds of rocks.

To discuss the final factor that matters to water chemistry, I have to switch my analogy from tea to coffee. Back in the days before gourmet coffee, you'd often get a really terrible cup of coffee at gas stations and diners, where they would make a pot first thing in the morning, then leave it in on a hot burner all day. Over the course of the day, the heat would concentrate the coffee, making it strong and bitter (this coffee was usually served in a Styrofoam cup, which complemented its vile flavor). Some lakes, including the alkali and salt lakes I've mentioned, have no outlet—water does not leave the lake by a surface stream or groundwater. As a result, materials entering these lakes are concentrated by evaporation just like the coffee in that gas station pot, albeit over thousands of years, producing distinctive water chemistries.

All of these factors—the contents of the watershed, contact time with materials in the watershed, and degree of concentration by evaporation—vary greatly from one place to another across the surface of Earth, producing the enormous range in water chemistry that we see in inland waters. Just as you can get pretty much any kind of tea you want by picking different tea bags and steeping times, inland-water organisms can find almost any kind of water chemistry they need by choosing among different kinds of watersheds.

Without dragging you element by element through the periodic table, I'll make the obvious point that inland waters differ enormously in almost every aspect of their chemistry, just as they do for pH. There are inland waters about as pure as distilled water, inland waters more than seven times as salty as the sea, and inland waters like China's Yellow River and its tributaries whose water is more than half sediment by weight, raising the question of whether they are inland waters or just wet soils running downhill.[7] There are inland waters where concentrations of essential nutrients like nitrogen and phosphorus are almost unmeasurable, starving their few inhabitants, and inland waters where nutrients are superabundant, fueling runaway growth of algae and plants. There are inland waters browner than a cup of tea as a result of the dissolved organic matter that they contain and some so perfectly clear that visibility exceeds 240 feet (74 meters).[8] There are inland waters containing harmful amounts of natural poisons like arsenic and selenium. And all of this offers a nearly infinite variety of chemical habitats for inland-water life.

The variation in water chemistry adds to the equally large variation in the physical characteristics of inland waters that I have described in earlier chapters, resulting in enormous variety among the world's millions of inland waters. Even though each of these bodies of water is made up of the same substance as the ocean (water, rather than, for example, liquid ammonia) subject to the same constraints of gravity, exposed to the same collection of chemical elements, and so on, inland waters are not simply millions of tiny clones of the ocean, each containing a diminished subset of the ocean's species—a few starfishes in one lake, a couple of oysters in another. Instead, using the same basic materials as were used to build ocean ecosystems, inland waters have appeared in an infinite number of variations on that oceanic theme. Before we go on to take a look at what kind of species do live in inland waters, we need to consider one other factor that has been important to the evolution of that diversity.

Chapter 7

ISOLATION

All Inland Waters Are Islands

When most people think of an island, they think of a sand-fringed tropical isle covered with palm trees and surrounded by turquoise waters, maybe with Gilligan, the Skipper, and Mary Ann (at least that's what comes up when you google "island").[1] But when biologists think of islands, they often think of isolated evolutionary laboratories that produce unique and interesting species. The Gálapagos are perhaps the most famous of these laboratories, with their finches and tortoises that have inspired biologists from Darwin forward. There are many other examples of distinctive and even bizarre species that live on one island or archipelago and nowhere else, some well known even to laypeople (the now-extinct dodo on the island of Mauritius) and some only known to evolutionary biologists (the more than 1,000 unique species of fruit flies on Hawaii, or the hundreds of species of flightless rails on Pacific islands, most living on just one or a few islands).[2]

Evolution can produce unique species on islands in several ways. Species may adapt to the local conditions on the island. Thus, many species of island birds lost the ability to fly because there were no predators on the island that they had to fly away from and because flight muscles take a lot of food to build and maintain. Island species may evolve to fill empty niches, which may be especially common on small or remote islands that contain few species. Often, the genetics of island populations may diverge from that of their ancestors simply by chance, either if the island population was founded by just a few individuals or because the size of the island keeps the population small. Finally, adaptive radiation may produce whole flocks of new species on islands, like the Hawaiian fruit flies. The result of these processes is that islands in oceans contain more than their share of biodiversity—islands up to the size of Greenland may constitute only 5% of global land area, but they contain 15–20% of the world's vascular plant, mammal, bird, and amphibian species.[3]

Not all islands are covered by palm trees and surrounded by water. In fact, every body of inland water is a kind of island, surrounded by stretches of dry land that may be as inhospitable to its inhabitants as the Pacific Ocean is to Gilligan and Mary Ann. Like islands in the sea, inland-water "islands" come in all shapes and sizes, from puddles to the North American Great Lakes. Although some lakes are shaped like familiar oceanic islands (irregular circles and ovals), many inland-water islands have weird shapes unlike any island in the ocean—spidery archipelagos consisting of all of the connected streams, lakes, and wetlands in a drainage basin.

The degree to which a lake or river system acts as an island depends not only on its geography and geometry but also on the species being considered. For a deepwater sculpin, a specialized fish that lives only in the deepest waters of lakes, the lake itself is the island. There is no way that it can get from the hospitable deep water of one lake to the next without crossing vast stretches of inhospitable shallow water or dry land. For a snail that lives on rocks in the lake shallows as well as in streams, the entire drainage system that contains the lake constitutes the island, but lakes and rivers in the adjacent river basin form another island remote from the first. For a dragonfly whose larvae live in the lake shallows but whose adults can fly for miles, the lakes are islands of suitable habitat, but neither the lake nor the river basin is an island in the sense of confinement, because it can easily fly to other lakes. In this respect, the dragonflies are like the oceanic birds that can visit islands all across the Pacific.

It may surprise you to learn that there probably are more of these watery islands floating in the sea of land than there are islands of land in the sea. The Global Islands Database lists 175,000 oceanic islands bigger than 25 acres (10 hectares), but there are about 2 million lakes of this size.[4] Then there are all of the river drainage systems on top of this. So when you think of a typical island, maybe you should think of a small lake in Sweden with the Arctic char that are stranded there rather than a tropical island with Gilligan and the skipper.

The fact that many inland-water organisms live in a world that consists of a series of more or less isolated islands has several important consequences for their evolution, ecology, and conservation. It means that evolution may proceed independently in each isolated body of water, sometimes producing different species in each river basin, lake, or spring, that many inland-water species have small geographic ranges (with the entire global population of the species being confined to a single river basin, lake, or in extreme cases even a single spring), and that they are very vulnerable to catastrophes, whether natural or (as is more common nowadays) caused by human activities. If an isolated population is destroyed by a catastrophe, it may be a long time before that habitat is recolonized by new colonists from another "island." Worse yet, if the isolated population that was destroyed was the only existing population of that species, then global extinction results. All of these effects are stronger for species that have trouble moving between islands (like the deepwater sculpin) than species that can move freely between islands (like the dragonfly). Chapter 13 looks more closely at just how strong these island effects have been for patterns of endangerment and extinction among inland-water species.

Chapter 8

LIFE
Inland-Water Biodiversity

We've established that all inland waters are islands, more or less isolated from other inland waters, and that physical and chemical conditions can vary radically from one "island" to the next. This combination of isolation and environmental diversity would seem ideal for allowing many species to evolve and sustain themselves in inland waters. Indeed, although they cover only about 1% of Earth's surface (depending on what you count as a fresh water), *fresh* waters support about 150,000 known animal species (the biological diversity in inland brackish and salt waters seems not to have been catalogued), which is about 10% of all animal species known from our planet, about half of all known fish species, and about a third of all known vertebrate species.[1] Approximately 2,600 species of freshwater plants are known.[2] Inland waters also support many microbes and algae, but the number of species is less well known; the number of inland-water algal species is probably in the tens of thousands.[3]

These figures are changing and imprecise because biologists still don't know how many species live in inland waters or indeed on land or in the ocean either. Biologists are still finding and naming new species all the time, even among well-studied and conspicuous animals like fishes. For instance, about 150 new species of North American inland-water fishes—a well-known group of animals in a well-studied region—have been described since 1990, and biologists know of at least 50 more species that are awaiting formal description and naming.[4] We know so little about many kinds of smaller organisms that we don't even have a good estimate of how many species remain to be found, but at least many thousands of nonmicrobial species living in inland waters are still waiting to be discovered. So stay tuned. And quite a few inland-water species probably went extinct as a result of human activities before biologists ever discovered them.

The catalogue of known freshwater species is dominated by insects (about 75,000 species), various kinds of algae (tens of thousands of species), fishes (approximately 15,000 species), crustaceans (12,000 species, including crayfishes, crabs, and prawns, as well as many tiny animals), amphibians (8,000 species, including frogs, toads, and salamanders), mites (which look like tiny spiders, 6,000 species), and mollusks (5,000 species, including snails, mussels, and clams).[5] The list of major biological groups living in the freshwater realm is strikingly different from that for the ocean, where insects scarcely occur, and land, where fishes are absent and few crustaceans are found (the pill bugs in your garden are a notable exception). On the other hand, echinoderms (starfishes, sand dollars, sea cucumbers, and their relatives) and cephalopods (squids and octopuses) are among the groups that are common and conspicuous in the ocean but missing from inland waters.

You may be surprised to learn that some kinds of animals that you think of as ocean dwellers also inhabit fresh waters. Sponges are common and widespread in fresh waters, although most freshwater species are small and inconspicuous.[6] There are a few freshwater jellyfishes, beautiful and delicate, none of them large enough to sting people.[7] When people think of cetaceans (whales, dolphins, and porpoises), they think of the blue ocean, but several species of dolphins and porpoises live in great rivers around the world, chiefly in the tropics.[8] Indeed, the first cetacean that went extinct owing to human causes was a freshwater species, the baiji, which appears to have been eliminated from China's Yangtze River around the year 2000.[9] Sharks even stray into some rivers and lakes, chiefly in the tropics, although it is doubtful whether there are any truly freshwater sharks (phew).[10]

Ecologists toss around biodiversity statistics all the time—so many species of this, so many species of that. But such summary statistics about the numbers of species, however impressive, can go only so far in describing inland-water biodiversity. When you look at the statistics of a great baseball player—home runs, runs batted in, earned run average, and so on—it's easy enough to tell that they were a hall-of-famer. But you don't really want to see statistics—you want to see Joe DiMaggio taking off at the crack of the bat to pluck a fly ball out of the outfield grass, the graceful rhythm of a double play, the long arc of a home run hit into the left-field stands.[11] In this and the next four chapters, I say a little more about the species that live in inland waters in the hopes that I can give you at least a glimpse of the ecological equivalent of a perfect double play.

AN INTRODUCTION TO INLAND-WATER LIFE

If you and I had a few hundred thousand pages to spend together, we could go through inland-water species one by one and get a really good picture of their diversity. Neither of us has the time for that, so I start with an abbreviated box-of-chocolates approach, looking at just a few species and concentrating our attention especially on the beautiful

Fig. 8.1. A small sample of the diversity of fishes found in inland waters. (*Top left*) the Amazon leaffish, which mimics a dead leaf (with a real dead leaf for comparison); (*Bottom*) the Syr Darya sturgeon, a small sturgeon—adults are only 1–2 feet (30–60 centimeters) long—from the Aral Sea and its tributaries in Central Asia, now extinct or nearly so; (*Top right*) two different species of armored catfish from Brazil with different paint jobs. *Sources: Top left, top right:* Mark Sabaj, Academy of Natural Sciences of Drexel University and iXingu Project, NSF DEB 1257813; *Bottom:* Karl Federovich Kessler, *Puteshestvye v' Turkestan* (St. Petersburg, 1874), Wikimedia Commons.

and the bizarre. In chapters 9 through 12, I go into a little more detail by describing some of the adaptations of inland-water species.

Probably the simplest way to appreciate the diversity of freshwater organisms is to look at pictures (figs. 8.1–8.8). Even if you know nothing about the physiology, behavior, life histories, or other adaptations

Fig. 8.2. A few remarkable inland-water reptiles and amphibians: (*Top left*) the Yangtze giant softshell turtle from China and Vietnam, which can exceed 500 pounds (227 kilograms) and is the largest inland-water turtle, now extinct or nearly so; (*Middle left*) the giant snake-necked turtle from Australia; (*Bottom left*) the Japanese giant salamander; (*Right*) the green anaconda, which some people say can reach 550 pounds (250 kilograms) in weight and more than 30 feet (9 meters) in length; regardless of the exact dimensions, it certainly gets larger than a snake needs to be. *Sources: Top left*: John Edward Gray and/or G. H. Ford, Wikimedia Commons; *Middle left*: Sam Fraser-Smith, CC BY 2.0, Wikimedia Commons; *Bottom left*: Naturalis Biodiversity Center, CC0, Wikimedia Commons; *Right*: Dick Culbert, Gibsons, BC, Canada, CC BY 2.0, Wikimedia Commons.

of these species, you can see from the pictures that inland waters contain a highly diverse array of species, spanning a huge range of sizes, shapes, and colors.

Let's start with the fishes, which are the first animals that most people think of when they think of aquatic life. You probably think that you're familiar with these fishes. After all, inland waters include some our most popular sport fishes (e.g., trouts, black basses, pikes), food fishes (e.g., carps, tilapias, catfishes, whitefishes), and aquarium fishes (e.g., tetras, guppies, goldfish). But these few dozen inland-water fishes that most of you know about are just a tiny fraction of the 15,000 known species in Earth's inland waters. Fig. 8.1 supplies a small sample showing how varied inland-water fishes can be. Inland waters support magnificent large fishes like the giant freshwater stingray of the Mekong (up to 660 pounds [300 kilograms]), the arapaima of the Amazon basin (up to 10 feet [3 meters] long and more than 400 pounds [180 kilograms]), and the toothy alligator gar of the American South (reaching 10 feet long and 350 pounds [160 kilograms]). At the other end of the spectrum the tiny *Paedocypris progenetica* (which has no common name) of Indonesian peat swamps is a contender for the smallest vertebrate on the planet, mature at a length of 0.3 inches (7.9 millimeters) and not known to get bigger than 0.4 inches long (10.5 millimeters).[12] Inland-water fishes come in all colors, from alarming bright colors and patterns to silver as bright as a new coin to subtle camouflages to nearly transparent (fig. 8.1). Shapes range from muscular, streamlined salmons to elongate pikes and eels, chubby bullheads and sculpins, and the pancake-shaped sunfishes. The habits and adaptations of inland-water fishes are as varied as their appearances.

There are plenty of vertebrates other than fishes in inland waters. More than three quarters of the world's turtle species live in inland waters (fig. 8.2, *top and middle left*), along with some excessively large snakes (fig. 8.2, *right*).[13] Even scarier than the snakes are the more than 20 species of crocodilians living in tropical and subtropical inland waters around the world.[14] These are the largest animals in inland waters, with several species reaching weights of more than 1,000 pounds (450 kilograms). Inland waters also provide habitat for most of the world's 8,000 species of amphibians, at least when they are young.

Inland-water amphibians include the world's largest (and perhaps ugliest) salamanders (fig. 8.2, *bottom left*), which can be more than 5 feet (1.5 meters) long. Hundreds of bird species like herons, ducks, gulls, darters, rails, and sandpipers depend completely on inland waters for food or habitat. One of the most interesting is the American dipper, which feeds as it walks along the bottom of swift streams. Several well-known mammals live in inland waters (platypuses, beavers, muskrats, capybaras, nutrias, Baikal and Caspian seals, otters, minks, manatees, hippopotamuses), as do a number of more obscure mammals whose names reveal their habitat preferences (water shrews, water opossums, otter shrews, fish-eating mice, swamp rats). Although just a few cetaceans live in inland waters, among them are some of the most interesting and imperiled species in inland waters. Of course, many other species of "terrestrial" vertebrates live along the margins of inland waters (deer, raccoons, salamanders, and flycatchers, waxwings, and other birds that feast on emerging aquatic insects, to name just a few) and depend on these waters for food, water, and habitat.[15]

DIGRESSION 8.1

A World of Catfishes

As an example of the proliferation of inland-water life, consider the catfishes. You might think that evolution would have produced a single all-purpose inland-water catfish, good for all occasions, that thrives in all kinds of waters. Indeed, there are catfishes that would seem to fit the bill. Channel catfish (fig. 8.3), for example, are beautiful, sleek catfishes that can reach about 50 pounds (23 kilograms). They live in creeks, rivers, lakes, and reservoirs and are even cultured in ponds (this is usually what you're eating when you order catfish in a restaurant in the United States). They thrive in all kinds of climates—their native range extends

Fig. 8.3. The channel catfish, a good, all-purpose catfish. *Source:* Smithsonian Environmental Research Center, CC BY 2.0, Wikimedia Commons.

from Mexico to Canada, and they've been successfully introduced around the world. They don't mind a little salt, so they do fine in mildly brackish lakes and estuaries. They'll eat almost anything that fits down their gullet, including fishes, crayfishes, insects, mussels and clams, snakes, ducklings and small mammals, plants, seeds, and (if offered as bait) rotten cheese, hot dogs, dough, and soap. Once you have this species, why would you need any other kind of catfish? It's not too hard to imagine a world in which inland waters were populated by this single, adaptable catfish species and no others.

Evolution had a different idea about catfishes, though, producing 3,000 known species (more are being discovered every year) in the world's inland waters.[16] Instead of there being a single, all-purpose catfish, there are catfishes for every occasion. There are tiny, cute catfishes no bigger than your fingers. There are catfishes weighing hundreds of pounds lurking beneath the surfaces of the world's great rivers (one of these huge catfishes, the European wels, has learned that pigeons are good to eat and swims right out of the water onto the riverbank to pluck unwary birds off of the beach, a "whiskered killer whale" in the language of the biologists who discovered this behavior).[17] There are slender catfishes and fat ones, catfishes with thin skins as smooth as

Fig. 8.4. A glass catfish (*Kryptopterus vitreolus*). *Source:* Conterally, CC BY 3.0, Wikimedia Commons.

silk and catfishes armored with bony plates. Some catfishes are blind and pale, living deep underground in caves and aquifers, while others ("glass catfish") are nearly transparent (fig. 8.4). Many catfishes can breathe air, and some can even "walk" across the land from pond to pond on rainy nights.

Not only did evolution produce many types of catfishes in inland waters but it also usually produced many species of each type. For instance, every continent (except Antarctica) has its own species of 100-pound-plus (50-kilogram-plus) catfish lurking in its rivers (fig. 8.5). In North America, we have the blue and flathead catfishes; in Europe, the wels; in Asia, the Mekong giant catfish (one of the world's largest freshwater fish, reaching weights of 600–800 pounds [270–360 kilograms])[18] and the goonch (which has a mouth filled with sharp teeth and is said

Fig. 8.5. Catfishes range from the titanic to the tiny. (*Left*) the piraiba from South America, which is said sometimes to eat people; (*Right*) the freckled madtom, one of dozens of species of finger-sized catfishes from eastern North America. *Sources: Left:* Eliodmsr, Wikimedia Commons; *Right:* Clinton and Charles Robertson, RAF Lakenheath, UK/ San Marcos, TX, CC BY 2.0, Wikimedia Commons.

to eat people); in South America, the piraiba (another reported human eater); and in Africa, the vundu and African sharptooth catfish. At the other end of the size spectrum, scientists have identified and named 29 species of madtoms (those cute, finger-sized catfishes) just in eastern North America (fig. 8.5) and suspect that several more remain to be named. A few of these madtoms have broad ranges extending over several states, but many live in just a single river system over a range of a few counties (one species, the Scioto madtom, was known to inhabit only a single series of riffles—shallow, fast-flowing parts of streams—in central Ohio before it apparently went extinct sometime after 1957).[19]

So it appears that evolution had an inordinate fondness for catfishes as well as beetles.[20] The large number of catfish species and the great diversity of form and function that these various species represent are matched to the extraordinarily varied habitats offered by the inland-water realm and encouraged by its fragmentation into countless island-like pieces. Although catfishes may be an outstanding example, this pattern of high diversity and small species ranges is characteristic of inland waters and is evident across many kinds of plants and animals.

Far more numerous than the vertebrates, both in terms of the numbers of species and the numbers of individuals, are the invertebrates, which include insects, clams, snails, crayfishes, shrimps, worms, and a bunch of other animals that you've probably never heard of (such as the gastrotrichs, horsehair worms, rotifers, and other obscurities that appear in the next few chapters). The invertebrates are also far more varied than the vertebrates; for example, the largest inland-water invertebrate weighs about 100 billion times as much as the smallest one.[21] As an invertebrate zoologist, it was hard for me to choose what photographs to show you, but fig. 8.6 shows just a handful of the most spectacular or beautiful inland-water invertebrates. We meet more inland-water invertebrates in chapters 9 through 12.

Then there are aquatic plants, which are conspicuous in most shallow, quiet inland waters. Although there are far fewer species of plants than animals in inland waters, they exhibit a wide range of sizes (fig.

Fig. 8.6. A few striking inland-water invertebrates. (*Top*) a freshwater sponge growing on a mussel in the Xingu River, Brazil; (*Middle*) the largest known inland-water insect, the giant dobsonfly (*Acanthacorydalis fruhstorferi*), from Southeast Asia; note the 1-centimeter (0.4-inch) scale bar in the lower left; (*Bottom*) the Tasmanian crayfish, which is the world's largest inland-water invertebrate. *Sources: Top*: Mark Sabaj, Academy of Natural Sciences of Drexel University and iXingu Project, NSF DEB 1257813; *Middle*: Natural History Museum, London, CC0, Wikimedia Commons; *Bottom*: Terry Mulhern, "A Giant Lobster, by Any Other Name," *Pursuit*, December 14, 2018, https://pursuit.unimelb.edu.au/articles/a-giant-lobster-by-any-other-name.

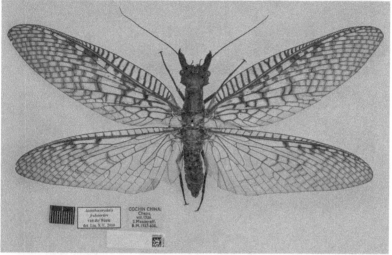

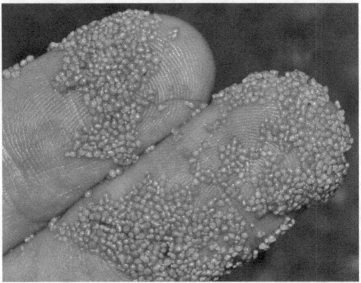

Fig. 8.7. Extremes in inland-water plants. (*Top*) the world's largest water lily, *Victoria*, which has the largest leaves of any plant native to the Amazon basin but is widely grown in botanical gardens, where it supports a small, weird industry of novelty photography; (*Bottom*) the world's smallest flowering plant, *Wolffia* (watermeal). A bouquet of a dozen of *Wolffia*'s flowers would fit on the head of a pin. *Sources: Top:* Missouri History Museum, Wikimedia Commons; *Bottom:* Christian Fischer, CC BY-SA 3.0, Wikimedia Commons.

8.7), forms, habits, and adaptations. Examples of such plants include water hyacinth and the tiny duckweeds (fig. 8.7, *bottom*) that float freely on the water's surface, rooted plants like water lilies whose leaves and flowers float on the water's surface (fig. 8.7, *top*), plants like cattails and bulrushes that stand with their roots in the water and their stems and leaves in the air, and many species that spend their lives underwater (although they may send their flowers up into the air). We meet more of these plants in later chapters.

Far less conspicuous but far more numerous than the aquatic plants, algae are the dominant primary producers in most inland waters (as well as in sea). Algae can live free-floating in the water (in which case they are called phytoplankton), attached to bottom sediments or rooted plants, or even inside of animals (as we see in chapter 11). A few inland-water algae are large enough to see without a microscope, although none is as large as the gigantic kelp of the seashore. For example, *Cladophora* is a filamentous green alga whose exuberant growths along the edges of nutrient-rich lakes and rivers look like waving masses of green hair (indeed, it is sometimes called mermaid's hair). It is eaten in Laos as a snack called kaipen.[22] Other large inland-water algae, called stoneworts, could be mistaken for rooted vascular plants by a casual observer. Although we normally think of plant and animal cells as microscopic, the individual cells of stoneworts can be as much as 6 inches (15 centimeters) long and so are favorite study subjects of cell biologists.[23] But only a few algal species are large enough to see with the naked eye, so you'll have to look under a microscope to appreciate the full beauty and diversity of inland-water algae (fig. 8.8). Among the notable groups of inland-water algae are the diatoms, which live in ornately patterned shells made of glass; desmids, which are beautiful bright green algae with elaborately symmetrical cells; dinoflagellates, which swim using their flagella; and cyanobacteria ("blue-green algae"), which can form noxious scums

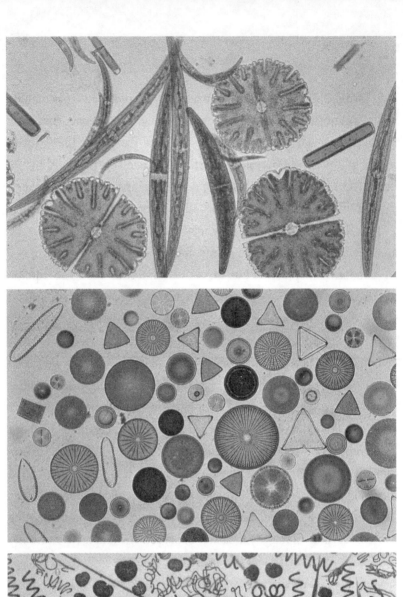
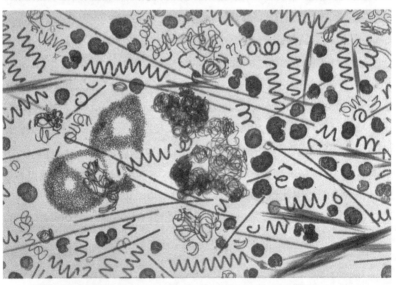

and produce potent toxins that can kill livestock, wildlife, pets, and people. Outbreaks of toxic cyanobacteria killed 76 people in a dialysis center in Brazil in 1996 and shut down the water supply for 450,000 people in and around Toledo, Ohio, in 2014.[24]

Finally, there are even smaller organisms in inland waters—protozoans, protists, fungi, bacteria, archaeans—that I lump together as "microbes" (see digression 2.1). We're in the midst of a great revolution in understanding microbial diversity because scientists finally have some good technological tools to study them. It now appears that ecosystems often contain hundreds or thousands (or more!) of kinds of microbes. Microbes often are very abundant in inland waters; bacteria typically occur in densities of around a billion per quart or liter of water or a trillion per pound (0.5 kilogram) of sediments, for example.[25] (I know what you're thinking—"A billion a quart—I'm never going to swim in the lake again!"—but almost none of these bacteria hurt people. They're a normal, healthy part of ecosystems, which would grind to a halt without them.) There's no point in showing you photographs of these bacteria, because they'd just look like little unremarkable blobs and dots. But their abilities are remarkable indeed. Various kinds of microbes can feed on almost any kind of organic matter (they can even use chemicals other than organic matter as food), thrive in boiling water, tolerate extremes of acidity, and prosper in oxygen-free environments. Although we won't be discussing this here, microbes also are vitally important in cycles of nutrients and other elements. Earth wouldn't be recognizable as Earth without the activities of microbes.

Fig. 8.8. A sampler of microscopic aquatic algae magnified to show their variety and beauty. (*Top*) desmids; (*Middle*) diatoms; (*Bottom*) cyanobacteria ("blue-green algae"). These diatoms are actually marine species but are quite similar to inland-water species. *Source:* Hilda Canter-Lund, Freshwater Biological Association.

Biodiversity Can Grow On You

When people quote the statistics about how many species live in fresh waters, they almost always leave out the parasites. But just as your pets get fleas, ticks, intestinal worms, and other disgusting parasites, inland-water species harbor their own array of parasites, which add to the diversity of inland-water ecosystems. This neglect of parasites is a little surprising. Although we don't normally think of biologists as squeamish, apparently there are degrees of squeamishness, and biologists who think nothing of handling snakes, eels, slimy snails, and all kinds of creepy many-legged bugs apparently recoil at the thought of parasites.

An astonishingly large part of the biodiversity on this planet consists of parasites. Scientists still don't have a good idea how many parasite species there are, but it has been seriously suggested that half (or more) of the world's animal species may be parasites.[26] Major groups of parasites in inland waters include worms such as flatworms, roundworms, spiny-headed worms, tapeworms, tongue worms, and horsehair worms; crustaceans like fish lice, copepods, and isopods; leeches; mites; a highly peculiar group of cnidarians called myxozoans that don't in the slightest resemble their cousins the corals, jellyfishes, and hydras; and many other species of microscopic invertebrates, protists, fungi, bacteria, and viruses. (Notice that I'm not showing you any pictures of these animals. You're welcome.)

Because of the stiff challenges of finding new hosts and outwitting the host's defenses, parasites have evolved some of the most remarkable adaptations of any species. We discuss the elaborate adaptations that help pearly mussels get their parasitic larvae onto fish in digression 9.1. Here are just two other examples. Amphipods are small cousins of shrimps that, like shrimps, are a preferred food of many fishes, birds, and other predators. Consequently, amphipods spend much of their lives hiding in dark places at the bottom of lakes and streams. After they are infected by the larvae of parasitic spiny-headed worms, which need to get from the amphipod into a fish, where the adult worms live, their behavior changes. Instead of wanting to go down, away from light, the infected amphipod becomes attracted to light, so it swims up off the bottom toward the light, where it presumably is promptly gobbled up by a fish.[27] Voilà, the parasite has reached its new host.

If you remember anything at all from the movie *Alien*, I bet it's the scene where that alien creature bursts out of John Hurt's chest. I'm sorry to have to tell you that such gruesome scenes are not just figments of the screenwriter's imagination but commonly occur in real life.[28] Both horsehair worms and mermithid roundworms have larvae that grow inside a host until they fill nearly the entire body cavity, then burst out of the host's body, which has become basically a living wrapper for the parasite. But here's a problem—the horsehair worms and some of the mermithids use terrestrial insects like crickets and grasshoppers as hosts for their larvae, even though their adults live in the water. So somehow, the parasite has to get from the land into the water. The parasites again solve this problem by hijacking the host's behavior. Infected insects are irresistibly drawn to water and upon reaching a body of water, the hapless insects throw themselves in, at which point the adult parasite bursts out of its host, neatly taken to its new habitat by a zombified host.

In cases where humans are the host, parasites are well adapted to infect people who touch, drink, live near, or eat food from inland waters.[29] Schistosomiasis is a terrible, debilitating (and sometimes fatal) disease that affects more than 200 million people in tropical regions around the world. The cause of this disease is a parasitic flatworm whose larvae swim from a freshwater snail (the intermediate host) into a human (the definitive host) who simply touches the water while bathing or wading. The adult flatworms live in the veins around the intestines, liver, and bladder, causing a wide range of harmful symptoms. The guinea worm is a roundworm that infects people who drink water containing tiny zoo-plankton infected with the larval worm. When it comes time to lay her eggs, the large female worm bursts through the skin, causing intense burning pain (many people think that the "fiery serpent" of the Old Testa-ment was a guinea worm).[30] People often try to relieve this pain by soak-ing the wound in water, at which point the female releases her eggs into the water. Guinea worm, which once affected tens of millions of people, is on the brink of being eradicated from the planet, a triumph of public health programs. But you don't have to have direct contact with water to get inland-water parasites. River blindness, which affects the vision of about a million people, is caused by roundworms that are transmitted to people by the bites of adult blackflies whose larvae live in nearby tropi-cal rivers. Finally, several common parasites can get into people who eat raw or poorly cooked inland-water fish or shellfish. This group contains some especially disgusting parasites, including the broad tapeworm, which grows to 30 feet (9 meters) long in the human intestine. Thus, in many parts of the world, human health is directly linked to the condition and biodiversity of inland-water ecosystems.

Because of problems like these, it's easy to think of parasites as simply being harmful. However, ecologists are now beginning to under-stand that parasites can play both positive and negative roles in eco-systems.[31] Parasitologists remind us that predators like wolves are cru-cial to stabilizing food webs and keeping prey populations from growing out of control (and that wolves and other predators were themselves reviled as purely evil just a few decades ago, when we tried to wipe them out). Ecologists are still working to understand the multiple, important roles that parasites play in ecosystems, but it is at least clear that elimi-nating parasites from ecosystems would not necessarily be a good idea for either the free-living species in the ecosystem or the humans that depend on those ecosystems.

I'm not sure that I can make you feel sorry for parasites, but we are now coming to understand that the conservation status of inland-water parasites is probably as bad or worse than that of free-living inland-water species, whose dire state I describe in chapter 13. Parasites are subject to a double jeopardy—they can disappear because they are sensitive to a threat (like climate change or pollution) or because their host is disappearing. As a result, there is a growing movement to include parasites in efforts to preserve Earth's biodiversity.[32] (Proponents of parasite conservation hasten to point out that they're not advocating for preservation of species that threaten other rare species or that might harm human health or livelihoods.)

Even though neither you nor most biologists may want to think about parasites, they are an interesting, important, and imperiled part of inland-water ecosystems and are worth careful management and conservation.

Here, I conclude this briefest of overviews of inland-water diversity. To underscore the point that diversity is not just about counting the number of species but also about the adaptations of these species, the roles that they play in ecosystems, and values they represent to people, I spend the next four chapters exploring functional consequences of that diversity. I focus on solutions to four problems that face inland-water organisms: getting washed down to the sea, losing their habitat when the water dries up (not a problem in the ocean!), finding food, and sex. These are just four of many possible examples I could have chosen, but I hope that they are enough to convey the extent of the highly diverse and amazing adaptations that allow inland-water organisms to thrive across the varied and at times stressful inland-water world.

Chapter 9

CHALLENGES 1
How Do You Keep from Getting
Washed to the Sea?

One of the few generalizations that applies to almost all inland waters of the world is that they contain some form of life and usually many species. Despite this obvious fact, inland-water biologists have spent a lot of time worrying about how inland-water species keep from getting washed to the sea. After all, except for rivers that run into closed basins like the Caspian Sea and the Great Salt Lake, all inland waters ultimately run downhill to the sea. Doesn't this mean that inland-water organisms should get washed downstream and eventually end up in the sea and that inland waters should eventually become depopulated, first in the torrential headwaters and then further downstream? (And for closed basins, you'd have the similar question: how do the species all keep from getting washed downhill to the Caspian Sea or the Great Salt Lake?) It is true that this constant downstream movement poses a challenge to some species, chiefly free-floating plankton, but the mere observation that even swift mountain brooks are full of life should tell us that organisms have solved this problem, even if biologists haven't. Let's look at a few of the ways that inland-water organisms are able to stay in place even in the strongest currents and are able to move upstream to keep from ending up in the deep blue sea.

If you're thinking of fishes when you're reading this, then you already know one answer—they can just swim upriver. Indeed, fishes and many other inland-water organisms are fine swimmers and can work their way upstream unless they are blocked by waterfalls, torrential currents, or a dry section of channel. Some fishes are truly magnificent swimmers and can migrate long distances against the current. The salmons are probably the most famous example. You may have seen pictures of salmons valiantly leaping waterfalls, and some salmons migrate 2,000 miles (3,200 kilometers) upriver from the sea to spawn.[1]

Although salmons are the best known of the migratory fishes, many other kinds of fishes migrate up into inland waters from the sea. Fishes such as sturgeons and shads are like salmons and migrate hundreds of miles upriver from the ocean to spawn in fresh water. Freshwater eels migrate in the reverse direction, departing inland waters as adults to spawn in the ocean, leaving their tiny young to migrate hundreds of miles against the current back into inland waters.

Far more numerous but less celebrated than the "diadromous" species that migrate between the sea and inland waters are the fishes that migrate within inland waters. These species typically migrate upstream to spawn, often traveling long distances within river systems. These riverine migrants, like their diadromous cousins, have been badly affected by the dams that block their migrations, which in turn can profoundly change ecosystems deprived of these migrants.[2]

Another way to move upriver is to get up out of the water and fly. Many aquatic insects spend most of their lives underwater, then emerge as winged adults to disperse and mate. Swarms of emerging insects can be so large that they show up as 50-mile-long (80-kilometer-long) clouds on weather radar and have to be cleaned up from lighted roads and parking lots with shovels or even snowplows (they are attracted to lights, so turn off your porch light during insect emergences).[3] Most of these insects fly only short distances, although long enough to get over obstacles like waterfalls, but there are exceptional long-distance flying migrants, too. The current champion among insect migrants, far outdistancing the famous monarch butterfly, is an inland-water dragonfly called the globe skimmer or wandering glider (*Pantala flavescens*), which probably can fly more than 1,500 miles (2,400 kilometers) without stopping and may migrate more than 3,700 miles (6,000 kilometers) in its lifetime.[4]

Finally, although it's not as cool as flying or even swimming, some species just walk upriver. You might think that walking would take animals no further than a mile or two, but some walkers cover impressive distances, especially considering their small body sizes. As far as I can tell, the mitten crab covers the longest distance on foot. Like eels, mitten crabs breed in the ocean, and the small juveniles then walk upriver, where they spend the next four or five years in fresh waters. During this brief life, they can travel more than 870 miles (1,400 kilometers) upriver, sometimes leaving the water to trundle around obstacles. Then they walk or drift all the way back down to the ocean to breed.[5]

Even snails undertake impressive migrations. Snails called neritids that breed in the ocean can migrate upstream more than 10 miles (16 kilometers) inland in enormous waves containing hundreds of thousands of snails and reach elevations of more than 1,200 feet (400 meters) above sea level.[6] Pretty impressive for an animal about a quarter inch (6 millimeters) long that can only crawl (yes) at a snail's pace.

So when I see a picture of a leaping salmon, as noble as it might be, I like to think of all of the other animals that don't have press agents as good as the salmon's that are fluttering, swimming, or crawling up the same stream without making such a fuss about it.

Taking the Express Uptown

You'd think that freshwater pearly mussels might have an especially hard time moving upstream—they move s o s l o w l y. There is a famously speedy mussel species that sometimes crawls as much as 3 feet (1 meter) per day, but when my colleagues and I tagged some mussels in a New York creek, we found that most of them moved less than 2 inches (5 centimeters) in 5 years. At that rate, the headwaters of a stream would be a many years' journey for a mussel, which all the time would keep getting washed back downstream by floods. Not only that, but the marine mussels from which freshwater species evolved have free-swimming larvae that spend days to weeks up in the water. Try that in a river, and your children will end up many miles downstream or back in the briny, briny sea.

Pearly mussels have managed to solve this problem and get around in running waters by tricking fishes into doing the hard work for them—they take the bus upriver. To start with, pearly mussels have dispensed with free-swimming larvae and instead almost all have nonswimming larvae that are fish parasites. The larvae attach to a fish for a few days to a few months, during which time the fish carries them around, sometimes over distances that would take untold generations for adult mussels to cover.

But if the pearly mussel larvae don't swim, how do they get onto the fish? This seems like a serious design flaw, especially considering that the larvae live for only a few days. What's worse, many mussel species are picky about the fishes that they use and will survive only on one or a few particular species of fishes. What are the odds that a mussel larva will bump into the right species of fish before its short life ends?

Not so bad, actually.

Fig. 9.1. Examples of pearly mussel lures: (*Top left*), (*Middle left*) "Trojan horse" packets of larvae of the Ouachita kidneyshell before and after they are broken open by a fish. The gray objects in the lower panel are individual larvae, ready to attach onto a fish; (*Bottom left*) a moving lure that is part of the mother's body and that resembles a minnow; (*Top right*) the moving lure (white spot) of the eastern pond-mussel doesn't look impressive to us, but it is very attractive to fishes; (*Bottom right*) the lure of the orange-nacre mucket that is attached to a clear "fishing line." The mother is visible just below the lure. *Sources: Top left, middle left, bottom right:* M. C. Barnhart; *Bottom left:* C. Ryan Hagerty, USFWS; *Top right:* Catherine Corey, Rhiannon Dowling, and David L. Strayer, "Display Behavior of *Ligumia* (Bivalvia: Unionidae)," *Northeastern Naturalist* 13 (2006): 319–332.

Evolution has provided pearly mussels with several kinds of remark-able adaptations that seduce fishes into carrying their larvae (fig. 9.1).[7] Here are a few examples. Rather than releasing individual larvae, some mussels package their larvae into little Trojan horses that look like fish food. Some of these packets even wriggle, making them yet more at-tractive to fishes. When a fish attacks one of these packets, expecting a tasty treat, the packet comes apart, releasing the individual mussel lar-vae (think of Ulysses and his men spilling from the original Trojan horse), which quickly attach to the fish. Gotcha.

These Trojan horse lures are convincing enough to fool fish biolo-gists as well as fishes. My friend Chris Barnhart took a vial of the Trojan horses of the Ouachita kidneyshell mussel (fig. 9.1, *top left*) to a scien-tific meeting, telling us that he needed help identifying some small fish that he had collected. We gullible PhDs all peered at the tiny "fish" and offered expert opinions about whether they were suckers, or minnows, or whatever. Nobody said, "Hey, wait a minute . . . these aren't even fish!" If we had been fishes, we all would have been wearing kidneyshell larvae that day.

In other mussel species, a part of the mother mussel's body forms a lure, usually a moving lure. These lures can be very realistic, closely resembling a small fish or a crayfish, for example. Even in cases where the moving lures don't look like much to us, they can be very attractive to fishes. When my student Catherine Corey was studying the modest lure of the eastern pondmussel (which just looks like a moving white dot; see fig. 9.1, *top right*), she put fishes into an aquarium with a displaying female mussel, and the fishes always attacked within minutes. When a displaying female is attacked, she instantly releases a cloud of larvae onto the attacking fish.

Again, these lures can be very convincing. Catherine and I once came across a displaying female of another species with a moving minnow-shaped lure in a clear New York river only about a foot deep and argued for about 2 minutes about whether it was a real fish (my opinion) or a lure (hers). It turned out to be a lure, so if I were a fish I would have been wearing lampmussel larvae that day.

A group of pearly mussels called the riffleshells lure fish in, catch them by the head, and release their larvae onto the captive fish before letting them go. The edge of the female's shell is armed with sharp little teeth (like the teeth on the jaws of a pair of pliers) that help to grip the wriggling fish.

Even more remarkably, some mother mussels make a lure out of their larvae, then play out the lure on a 3-foot-long (1-meter-long) clear fishing line (a long mucous strand made by the female), allowing the lure to swim in the current, just like the lures that human anglers use.[8] The intended target of these lures appear to be fish-eating species such as black basses and pickerels.

These are just a few of the lures that mussels use so that they can take the express upriver instead of having to walk. Scientists are still discovering and investigating hitherto unknown kinds of lures. Some scientists even think that mussels may have lures that use scent rather than sight to attract hosts such as catfishes.[9]

The life history that I've just described applies only to pearly mussels. A few other kinds of mussels in inland waters still use free-swimming larvae like their marine ancestors, and they are spectacularly successful in the modern world. A prime example is the zebra mussel, which now lives all over Europe and North America, where it is one of the most abundant freshwater animals and is still expanding its range. How does it succeed? First, it thrives especially in lakes, where its larvae can swim and grow without being washed away. But this still raises the questions as to how it got into all those lakes across two continents in the first place and how it survives in many rivers as well. The answer is that careless people have done for zebra mussels what fishes do for the pearly mussels. We spread them up and down rivers, from lake to lake, and even across continents in ballast water, on recreational boats and trailers, and through canals. Once they are established in a lake or somewhere upriver, their larvae easily spread to downstream waters. (You might think that I'm blaming people unfairly, because birds, turtles, and other creatures must unwittingly move zebra mussels around when mussels attach to bird feet or turtle shells. Maybe, but for the 18,000 years after glacial ice left Europe, birds and turtles did not move these mussels across Europe. It was only after people built canals linking central and western Europe with the Black and Caspian Sea regions, the native range of the zebra mussel, that they began appearing all across Europe.)[10] So once again we have an example where bivalves take the express upriver, relying on some unsuspecting creature to do the hard work instead of moving upstream under their own limited power.

In addition to being able to move upriver against the flow of the water, many inland-water species have adapted to life in flowing waters by evolving ways to hold their place tenaciously, even in strong currents. My favorite example, and one of my favorite plants, is riverweed. I am not graceful or athletic, so I've spent a lot of time nearly falling into streams—dancing frantically on unstable rocks, sliding down slick riverbanks, and bracing myself against strong currents. More than once (ok, about a hundred times), I've found myself wading in deeper and swifter water than I meant to be in as I tried to get a good sample or to a better fishing spot and wondering how I'm going to explain this to St. Peter in a few minutes when I am swept away and drowned. Just as I'm about to lose my footing, I find that the river bottom has changed from slick, tippy stones to what feels like grippy indoor-outdoor carpeting, and I am saved from drowning for long enough to recover some good sense and retreat to a safer spot in the river.

The lifesaver is hornleaf riverweed, a plant that grows in the strongest currents in rivers in the eastern United States. (Ecologists think that it can live only where currents are too strong for crayfishes and other animals that like to munch on riverweed). Its strong roots grow over and tightly grip the rocks of the stream bottom, and its leaves are tough enough to withstand buffeting by the current. Biologists aren't entirely sure how riverweed establishes itself amid the fast currents, but two methods are probably important.[11] The tiny seeds of riverweed are coated in a strong glue. When wetted and then dried, this glue sticks seeds firmly to rocks and is able to withstand currents of at least 7 feet (2 meters) per second. When the seed germinates, it immediately sends out a tiny rootlet that attaches to the rock, using an adhesive that it secretes. New riverweed plants probably are started from seeds during periods of low water when parts of the river rapids are exposed to the air. Alternatively, it appears that small root fragments that break off a plant and drift downriver may start a new plant when they become lodged between rocks. As long as the river bottom itself stays in place, these adaptations allow riverweed to colonize and grow in swift waters without being washed downstream. What's more, I suspect that riverweed actually stabilizes the river bottom by knitting together the cobbles of the river bottom into a big, tough mat.

I don't recommend that you go out and walk on riverweed, which is probably not great for the plant, but it is worth admiring if you're snorkeling or kayaking in a swift, clear stream in eastern North America. If you really want to see riverweeds, though, you need to go to tropical or subtropical areas, which contain about 300 species (only a single species, by contrast, lives in the United States).[12] Some of these tropical species live on waterfalls, turning the whole waterfall green.

Another species from Colombia is the most remarkable bright red, giving the impression that someone like Christo covered the stream bottom in fuchsia carpeting.[13]

Stream animals are pretty good about holding their place in swift streams, too, using several tricks to keep from being swept downstream.[14] If you've spent much time along (or in) a fast-moving stream, you've probably noticed that there usually are quiet spots downstream of rocks or logs, along the banks, or right down next to the sediments. Stream-dwelling animals know all about these refuges. Fishes may lie comfortably in the quiet water next to a boulder, then dart into the current for an instant to capture a morsel that is floating by. The layer of quiet water next to the stream bottom is too thin to help fishes— usually far less than $1/8$ inch (3 millimeters) thick—but thick enough for tiny insects and other creatures to live in a swift stream without being exposed to the brutal forces of the current.

Stream-dwelling animals have several ways of hanging on to the stream bottom in swift currents. Aquatic insects have claws for hanging onto stones, algae, or plants, and some insects tether themselves to rocks using silk threads. The larvae of net-winged midges probably are the champions for hanging on.[15] These tiny animals, which can cling even to the vertical faces of waterfalls, have six suction cups on their bellies, which they can engage or release at will. Think of the fancy gloves that Tom Cruise used to climb the glass exterior of the Burj Khalifa in *Mission Impossible: Ghost Protocol* but with much better technology. While Mr. Cruise struggled to keep from falling, the midge larva can withstand forces of about 1,000 times its own body weight. Eat your heart out, Ethan Hunt.

Some animals hold their places in the current using the same devices that race-car drivers use to keep their cars on the track. Because current speeds are higher above a bottom-dwelling animal than under

Fig. 9.2. The northern hogsucker. The Committee for Replacing Unfortunate Names, which I just now made up, has recommended renaming this fish the "magnificent riffle fish" to recognize its elegant adaptations to life in swift waters—streamlined body, steep forehead that counteracts lift forces, a flat bottom and large pectoral fins for lots of contact with the sediments, cobble-and-sand camouflage, and an extendible mouth to hoover up delicious food from the sediments. *Source:* Brian Gratwicke, CC BY 2.0, Wikimedia Commons.

it, hydrodynamic lift tends to pull the animal up off of the stream bottom in the same way that aerodynamic forces lift a race car off the track.[16] To counteract these lift forces, both race cars and stream animals have low, steeply angled hoods. The drag forces that push animals downstream and slow down race cars can be reduced by streamlining. Finally, high friction, provided in fishes by a flat-bottomed body and big fins that press against the stream bottom and in race cars by soft, treadless tires, keeps fishes and race cars from slipping. One fish biologist has even claimed that a *dead* hogsucker (a kind of fish that lives in swift waters [fig. 9.2]) will hold its place in a riffle if carefully placed.[17] (I know, a normal person, if given a dead hogsucker, would say "ick" and back away instead of experimenting with the carcass, but fish biologists are not always normal people.)

Despite the many adaptations that allow inland-water species to hold their place in running waters, it is true that planktonic organisms, which live free-floating in the water, are constantly lost downstream and may have trouble maintaining populations in streams and rivers. Consequently, plankton is scarce or missing from swift headwater streams and during periods of high flows in all streams and rivers. It may surprise you, though, to learn that plankton is common during periods of low flow in many rivers, apparently constantly replenished from lakes, marshes, and quiet backwaters upstream, and augmented by growth and reproduction in the river itself. Even free-floating organisms, apparently so vulnerable and helpless against the forces of downriver flow, have been able to solve the challenge of downstream losses that has so fascinated biologists.

Writing about the problems that free-floating organisms have with being washed out of flowing waters reminds me about a final way to stay in place in streams and rivers—get rid of life-stages that live up in the water! In the ocean, all kinds of invertebrates—worms, snails and clams, various crustaceans—have tiny, free-floating larvae that drift with currents for days or weeks to find new homes. If you tried this sort of thing in a river, your new home would be in the sea, not in the river. Biologists noticed a long time ago that free-floating larvae are scarce among inland-water species (but see digression 9.1 for an interesting exception).[18] Instead of first appearing as tiny, floating larvae, most inland-water invertebrates are released as sticky eggs or well-developed juveniles (some fingernail clams even give birth to young that are sexually mature, which would be like giving birth to a teenager, if you can imagine). Some biologists have argued that the primary benefit of losing a larval stage is not to avoid being washed downstream but instead to aid in ionic balance (osmoregulation). Tiny larvae have high surface-to-volume ratios, so it would be hard for them to maintain their ionic balance in fresh waters, which are far more dilute than ocean waters.[19] Presumably, dispensing with a larval stage has both prevented inland-water invertebrates from being washed downstream and aided in ionic balance.

So there are many ways that inland-water species can fight the inexorable push of water flow to the sea and make their way upstream, whether by swimming, flying, walking, tricking someone else into carrying them, gluing themselves to the stream bottom, or dispensing with free-floating life stages. Now let's move on to a real challenge for all aquatic species—what do you do when the water dries up?

Chapter 10

CHALLENGES 2
What Do You Do When the Water Dries Up?

Lake Baikal and other ancient inland waters have existed continuously for millions of years. But many other inland waters dry out, some predictably every dry season and others capriciously at irregular intervals. Indeed, many "lakes," "streams," and "wetlands" spend more time dry than wet. If bodies of inland waters are islands in a sea of land, then these dry periods would be like Gilligan's island disappearing beneath the waves from time to time. Gilligan and his hapless shipmates would have been goners. How do the organisms that live in inland waters manage when their habitat disappears? Let's look briefly at four solutions to this problem.

The first and most obvious solution is to leave before the water disappears and move to a place where there is still water. If the drying habitat is connected to other waters, as in the case of drying streams and floodplain pools, for example, then animals can swim, drift, or walk into areas that still contain water. However, animals like snails, mussels, and many other small creatures move too slowly to be able to escape easily and even mobile animals like fishes must not wait too long before moving to more permanent waters.[1] Stay and hope that a summer thunderstorm replenishes the stream, or wait and die from suffocation, heat, desiccation, or predation as the stream shrinks away?

Flying insects don't require a watery escape route and can move to a nearby habitat that still contains water if it is within flying range and if they can find it. Furthermore, flying insects can leave at any time, so they can wait until the very last minute to escape from a drying stream or pond. A population of predaceous diving beetles living in a temporary livestock pond in Arizona offers a striking example.[2] The "pond" was dry for a year, then contained water for just 19 days after a rainstorm. At 10:30 in the morning on day 19, when the pond had shrunk to just 30 feet (9 meters) across and half an inch (1 centimeter) deep, the beetles "produced an intense, high-pitched buzzing sound" (I don't speak much beetle, but I'm guessing that this means "We're outta here"), rose up into the air in a single swarm, and headed southwest. An hour later, the pond had disappeared. Talk about cutting it close!

Many aquatic insects have life cycles that are synchronized to the annual cycle of wetting and drying of temporary aquatic habitats, whereby they grow wings and then escape before the water disappears. However, if insects are still immature (and wingless) when the water starts to disappear, they can be trapped in a drying pond along with all of the other nonflying species. There is some evidence that declining water levels can speed up development, allowing aquatic insects to earn their wings a little early and to escape from their disappearing habitat.[3]

Second, some inland-water organisms can stay in the dried-out habitat in a dormant state that is more or less resistant to drying. It is perhaps not too surprising that clams and mussels can close their shells tightly when the water disappears and survive for a while. Some can live out of water for only a day or two, but species that live in temporary waters can last a good long while. One mussel that lives in African temporary waters and goes by the local name of "hebabaeki" (which means "surprise") can live for months buried in the dried-out mud of a stream bed.[4] One specimen was taken from Tanzania to the British Museum in London, then placed into water after being kept dry for 12 months. To the surprise of the biologists, the mussel promptly woke up and tried to burrow into the bottom of the aquarium (and imagine the surprise of the mussel, which went to sleep in Tanzania and woke up in London).

Even a few fishes, which we think of as the most water dependent of animals, can go dormant when the water disappears. The African lungfish (fig. 10.1) is a remarkable example.[5] These fish, which as their name suggests have lungs and breathe air, burrow into the mud and form cocoons when the water dries up. During this dormant period, which can last for more than a year, their metabolism slows to less than 5% of normal (which helps in preventing starvation when you're not going to eat for a year) and their urine changes from ammonia to urea (which is much less toxic than ammonia and can safely build up in the blood). Because the sleeping fish continues to breathe air, the lack of water isn't a problem for respiration, as it would be for most fishes or other creatures with gills. The cocoon itself, initially thought to be simply an envelope to keep the fish moist, turns out to be a living organ whose powerful antimicrobial activity protects the sleeping fish from harmful bacteria. If the cocoon is removed or damaged, the dormant lungfish quickly succumbs to bacterial infection. The example of the lungfish shows that for an aquatic organism to use long-term dormancy to survive in temporary waters, it must not just avoid desiccation and suffocation when exposed to air but also possess a whole array of coordinated adaptations.

Perhaps the supreme examples of the long-term sleepers are bdelloid rotifers and water bears (also known as tardigrades), both tiny invertebrates that live in mosses and other semiaquatic habitats that

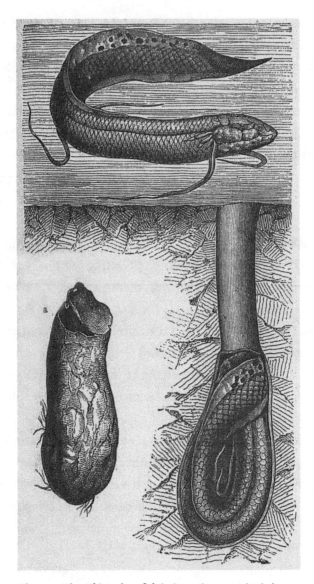

Fig. 10.1. The African lungfish in its active state *(top)*, its cocoon *(lower left)*, and a cross-section of the cocoon showing the dormant fish *(lower right)*. *Source: Meyers Konversations-Lexikon: Eine Encyklopädie des Allgemeinen Wissens,* 4th ed. (Leipzig: Verlag des Bibliographischen Instituts, 1891), 18:287.

are prone to dry frequently.[6] When confronted with drying, these animals gradually shrivel up (fig. 10.2), losing water until they reach a water content as low as 1% of normal, shut down their life processes so completely that signs of life are nearly undetectable, and go dormant. G. Evelyn Hutchinson, a great inland-water ecologist who had a way with words, referred to these animals as being "reversibly dead."[7] Remarkably, they revive promptly when water returns. Biologists are still trying to understand the biochemical and physiological adaptations that make this possible. It's not known how long tardigrades and bdelloids can survive in their deeply dormant state: it is well documented that they can last for 5 or 10 years, and there are repeated but unconfirmed reports of much longer survival.

DIGRESSION 10.1

Water Bears in Space

If drying out doesn't kill dormant water bears, what does? Since biologists first discovered that water bears are so resistant to drying, a small, strange subspeciality in biology has devoted itself to finding out exactly what *does* kill them. Experiments have now shown that water bears can survive boiling water, temperatures of 460 degrees below zero Fahrenheit (-273 Celsius, less than 0.1 Celsius above absolute zero), and pressures ranging from almost a total vacuum to 600 atmospheres (60 megapascals, 6 times as great as the pressure at the bottom of the deepest part of the ocean). They can endure being shot in a bullet at speeds of 2,000 miles per hour (3,200 kilometers per hour), gassed with methyl bromide, or exposed to gamma radiation of 5,000 grays (which equals 500,000 rads, about 1,000 times the lethal dose for you and me). They can stay alive for at least 30 years frozen at 4 degrees Fahrenheit (-16 Celsius) and survive shocks of 1.14 gigapascals.[8] When you read through this odd corner of the scientific literature, it's hard not to think of those Saturday morning cartoons where Wile E. Coyote tries to kill the roadrunner using a series of bizarrely specialized devices from the Acme Company.[9] I almost expect to open a scientific journal some day and see an article titled "The Effects of Falling Anvils on the Water Bear *Milnesium tardigradum.*"

It seems natural, then, that one of these biologists would call up the European Space Agency one day and ask them if they could shoot some tardigrades into outer space ("Yes, this is, uh, Professor Coyote, and I'd like to propose an experiment"). The agency agreed and sent two species of water bears up to experience the icy vacuum of space for ten days. To be sure, some of the water bears died (apparently, the intense ultraviolet radiation from the sun, 1,000 times stronger than what we experience down on Earth's surface, was tough even for water bears), but some of the animals survived even this ordeal, crawling away when rewetted back home on Earth and presumably muttering, "I just had the strangest dream."[10]

Third, instead of going dormant themselves, many inland-water species produce offspring that are highly resistant to drying.[11] Although they go by a variety of names (e.g., gemmules, resting eggs, opsiblastic eggs, statoblasts, ephippia, seeds), these resistant reproductive structures are basically seeds, except that they can sprout into animals as well as plants. Many kinds of inland-water plants and animals can produce such structures, which can protect the species not just against drying but also against many kinds of unfavorable environmental conditions that may occur in inland waters. As you might expect, such resistant structures are quite rare among animals that live in the ocean, which never dries up and where environmental conditions are more stable.

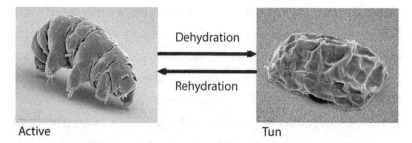

Active   Tun

Fig. 10.2. Scanning electron micrographs of active and dormant water bears. *Source:* Elham Schokraie, Agnes Hotz-Wagenblatt, Uwe Warnken, Brahim Mali, Marcuse Frohme, Frank Förster, Thomas Dandekar et al., "Proteomic Analysis of Tardigrades: Towards a Better Understanding of Molecular Mechanisms by Anhydrobiotic Organisms," *PLoS One* 5, no. 3 (2010): e9502.

Many of these dormant structures can last for more than a year, and quite a few have been found to survive for decades to centuries. The current record for longevity for an inland-water organism seems to be the 1,300-year-old seeds of sacred lotus that sprouted after being recovered from a dry lake bed in China.[12]

Furthermore, many of the long-lasting "seeds" of inland-water organisms are equipped with hooks or other structures that increase the possibility that they will be dispersed into another body of water by birds or other means. Between their long life and their dispersibility, there is a good chance that these resistant, dormant "seeds" left in a drying body of water will encounter favorable environmental conditions, either at another time or in another place.

The fourth and final possibility is just to die when the water disappears. This doesn't seem like much of a strategy if you're thinking about an individual organism (isn't the game over when you die?), but is not necessarily a disastrous outcome if you're thinking about the survival of a gene or a species. If the species can come back quickly from a refuge when the water reappears, temporary loss of a population at one or even many sites isn't such a disaster. Indeed, many of the insects that live in temporary waters are strong flyers and are good at finding water (like the predaceous diving beetle that found that temporary livestock pond in Arizona the first day it filled with water). Further, many have rapid life cycles, so that a generation or two may mature even in a short wetted period; 19 days was enough for that predaceous diving beetle to lay its eggs, hatch into larvae, mature into a new generation of adults, and fly away from that Arizona pond, and other temporary-water insects are even faster.[13] And smaller organisms like tiny winged insects, algae, and protozoans may be blown passively across the landscape, fall with rainwater, or be carried on birds and mammals into newly filled ponds and streams.[14] As long as a species can stay alive somewhere in the region and disperse into newly filled waters, then the death of an individual animal or even an entire population in a single drying water body isn't necessarily a terminal condition for the species.

DIGRESSION 10.2

If Inland-Water Species Have Superpowers, Why Do We Need to Worry about Hurting Them?

In this chapter and elsewhere, I've emphasized that some inland-water species have almost unbelievable abilities to survive challenging conditions. There are microbes living in lakes more acidic than battery acid (chapter 6), net-winged midge larvae clinging to the vertical faces of waterfalls (chapter 9), lungfishes living in bodies of water that are completely dry for months at a time, and tardigrades that survive, well, seemingly almost anything. You might be excused in thinking that human impacts couldn't hurt such tough organisms. Surely they can survive a little pollution or a short dry spell in a river that is drawn down by irrigation withdrawals. After all, Superman doesn't need to wear a seatbelt.

But as we see in chapter 14, a wide range of human activities have harmed inland-water species, to the point that many species are globally imperiled or extinct. How can this be?

The key is that *some* inland-water species have developed *some* powers to withstand remarkably stressful conditions in the habitats in which they have evolved. But the vast majority of inland-water species do not have such abilities, nor are the species able to deal with one particular problem necessarily able to deal with all problems (except perhaps in the case of tardigrades, whose broad superpowers are not yet understood). Think of it like this—Glenn Gould could play the *Goldberg Variations* brilliantly, and Usain Bolt could run 100 meters in 9.58 seconds, but I doubt that Gould was much of sprinter nor Bolt a wonderful pianist. And only a very few people can make a living as a professional musician or athlete.

The basic rule to remember is if we change a habitat in a way that subjects a species to a condition that is outside of its recent evolutionary experience, we run the risk of causing harm. Some of these changes might seem innocuous or even beneficial—eliminating the raging spring flood with a flood-control dam is good, right? But that spring flood may have cued fish that it is time to swim upstream to spawn, cleansed the river gravels of fine sediments so fish eggs could survive, or provided a nursery for the baby fish by inundating the floodplain. So when we pollute a lake or draw dry a formerly perennial river, we expose its inhabitants to stressful conditions they are not equipped to handle, often with deadly consequences.

For organisms that live in the eternal ocean, the prospect of coping with inland-water habitats that are so unreliable—freezing solid, getting torn up by violent floods, and even drying up entirely—must seem like an impossible challenge. The sudden drying of the ocean would be a disaster for most marine life, which would simply die.

While it is true that *most* inland-water species cannot survive their habitat drying out, even this brief account shows that evolution has produced a wide array of solutions to the daunting problem of ephemerality and that many inland-water species do just fine in "waters" that dry up from time to time. In fact, the first thing that a visitor to an ephemeral pond often notices is the *abundance* of life in the pond—the water is just filled with insects and fairy shrimps and tadpoles. Species have evolved to deal with freezing, flooding, and other apparently difficult challenges in inland waters through similar kinds of adaptations.

The existence of ephemeral waters adds greatly to the diversity and richness of the inland-water biota. Distinctive species live in temporary waters, and different kinds of species live in different kinds of temporary waters—streams, ponds, waters that dry for a month every year, waters that dry for 10 months every year, waters that dry regularly like clockwork, waters that dry and fill capriciously, and so on. Furthermore, different regions each have their own temporary-water species—Africa has four lungfish species, South America has another, and Australia has yet another (as well as a salamanderfish that acts a lot like a lungfish).[15] Just as for the highly acid volcanic lakes, distinctive habitats lead to distinctive groups of species in the highly diverse world of inland waters.

Chapter 11

CHALLENGES 3
How Do You Find Some Lunch?

Although most of us think about food a lot—just look at the number of cooking shows, foodie magazines, and cookbooks—we really have very parochial ideas about how to get food. For many of us, food comes from grocery stores or restaurants. In developed countries at least, only a few people grow their own food, and fewer still find our food in the wild. But inland-water animals (and some plants!) have had to be much more inventive when it comes to finding food.

Let's begin with some familiar ways of dining. Just as cattle and rabbits living up in our world feed on plants, aquatic animals browse on underwater plants. These browsers range from large animals like ducks and nutrias to minute insects that chew up leaves or burrow in tiny tunnels between the upper and lower surfaces of leaves. Even smaller animals feed on aquatic plants: microscopic roundworms poke tiny spears into plant cells and pump out their juicy contents.

In our terrestrial world, rooted plants are the most important primary producers, and they live nearly everywhere. In contrast, rooted plants are missing from many inland-water habitats, and tiny algae (fig. 8.8) live in many more places and account for a much larger fraction of primary production. Although these algae are mostly too small to see without a microscope, you may have noticed them as the slippery film on stream rocks or the scum on underwater branches or plant stems. These attached algae make good food (often better than aquatic plants, because they contain more nutrients and less indigestible material than the plants) and are grazed by what you might think of as miniature versions of cattle.

Many stream insects like mayflies and caddisflies feed on these films and scums that contain a blend of algae, bacteria, mucus and decaying organic matter, and bits of silt and clay, digesting the good parts. (Ecologists call this mixed material "biofilms," which shows you that ecologists would get nowhere in the restaurant business. If fancy restaurants served biofilm, they'd call it "mélange du ruisseau" or "fresh petites légumes in artisanal organic aspic.") Even smaller animals, scarcely larger than the algae themselves, pluck individual algal cells from biofilms and sediments the way that you pick strawberries from a field. Water bears and roundworms spear individual algal cells to suck out their contents.

In addition to this food that is rooted or attached, aquatic ecosystems contain bits of food floating in the water (plankton and bits of decaying organic matter), unlike anything we're familiar with up here in our world. It's as if Skittles and Snickers bars were floating through the air, just waiting to be gathered.[1] Many aquatic consumers have developed ways of capturing these tasty morsels of free-floating food.

Some animals live up in the water with the floating particles that are their food. Tiny animals like rotifers and *Daphnia* pluck individual bits of food from the water or pump water through delicate filters to capture food. Plankton-feeding fishes have structures called gill rakers on their gills that allow them to strain small plankton from the water.

But animals that live in the open water and feed on plankton are vulnerable to being eaten themselves because there is nowhere to

Fig. 11.1. Stream-dwelling insects that have special structures for capturing bits of food that float by. *(Left)* the head fans of a blackfly larva; *(Right)* a larval caddisfly with its silk net. *Sources: Left:* Douglas A. Craig, Douglas C. Currie, Leonardo H. Gil-Azevedo, and John K. Moulton, "*Austrocnephia*, New Genus, for Five Species of '*Paracnephia*' (Diptera: Simuliidae), with a Key to Australian Black Fly Genera," *Zootaxa* 4627 (2019): 1–92; *Right:* David H. Funk.

hide in the open water. To avoid becoming a lunch entrée, the little open-water animals often are transparent, or come up into the surface waters only during the night, hiding in the dimly lit depths during the day. Even the fishes of the open water are transparent or silvery and hard to see.

Other plankton-feeding animals live on stream or lake bottoms, waiting for currents and gravity to bring their food to them. Like their cousins in the water column, these animals use an ingenious array of pumps, fans, and filters to capture bits of floating food (fig. 11.1). My favorites are the net-spinning caddisflies, which build exquisitely perfect silk nets to catch tasty tidbits floating down the stream. While the net does the hard work, the caddisflies sit comfortably in their retreats, where I imagine them watching TV, or reading a trashy paperback, or doing whatever caddisflies do when they're not working.

The underwater world is also filled with predators that feed on living animal flesh—the lions and tigers of the underwater world. We all know about the predators that are so big that they menace people

(at least in our imaginations). Inland waters contain plenty of examples of these big, frightening predators—alligators (which really
do sometimes attack and kill people) and anacondas (the largest living snake species, sometimes more than 20 feet [6 meters] long and
more than 200 pounds [90 kilograms]—yeeek) and piranhas (yes,
they have killed people, but mostly they bite feet and hands). But the
inland-water world is also filled with smaller predators that are the
things of nightmares (or would be if you were a sixteenth of an inch
[1.5 millimeters] long).

Let's begin our brief tour of the tiny lions of the inland waters with
some that you might have heard of. Everyone loves the dragonflies
that patrol our wetlands and fields, and garden catalogues are filled
with art that features these elegant creatures. But before they develop
into adults, dragonflies spend months underwater as squat nymphs
that terrorize little creatures. The business end of a dragonfly nymph
is a masklike structure called a labium, armed with hooks and spines,
which fits onto the underside of the face. When a hapless little animal
approaches, the lurking nymph shoots out its labium to grab the prey.
The strike takes less than a tenth of a second, so unwary prey has little
chance of escape.[2] A pouncing cougar is a slowpoke by comparison.
Dragonflies have a liberal idea of what constitutes a good lunch; although they eat a lot of insects and zooplankton, they will also eat
tadpoles and fishes, and even have been seen reaching out of the water
to snatch up small frogs.

(When I was living at a biological field station one summer, we
had an aquarium in the living room. People who went out collect-
ing for a class or research would bring back some creatures for the
aquarium—insects, salamanders, tadpoles, zooplankton, little fishes.
By the end of the summer, everything had been eaten except for two
big dragonfly nymphs. Then one day, there was just one—the winner
of this sudden-death version of *Survivor*.)

If modern dragonfly nymphs about an inch (2.5 centimeters) long can kill and eat fishes and amphibians longer than they are, think about what griffenfly nymphs must have eaten. Griffenflies were ancient cousins of dragonflies that inhabited inland waters about 300 million years ago.[3] They were the largest insects that ever lived, the adults looking like enormous dragonflies, with wingspans up to 28 inches (71 centimeters). The nymphs, broadly similar to dragonfly nymphs, must have been close to a foot (30 centimeters) long. I haven't seen any studies of the diets of griffenfly nymphs, but they presumably attacked pretty much anything smaller than a breadbox.

Like dragonflies, the stoneflies whose adults are familiar to trout anglers spend most of their lives underwater. Some of these juvenile stoneflies feed on decaying leaves, but others are among the most ferocious predators in streams. They are so feared that even the smell of stoneflies in the water will cause mayflies (a favored stonefly lunch item) to hide away. Mayflies can spend so much of their time hiding from stoneflies that it reduces the amount of food that they eat, their growth rates, and the number of eggs that they lay.[4] (If your neighborhood were full of cougars and grizzly bears, you'd think twice about running out to get a pizza, too.)

Like packs of wolves, some inland-water predators work cooperatively to capture prey. Plenty of predatory fishes work in schools to round up and capture small fishes. Perhaps the most striking example of cooperative hunting by inland-water predators, though, is that across two species. In the Irrawaddy River in Myanmar, the local Irrawaddy dolphin population (now nearly wiped out from in-channel mining, fishing, and habitat destruction) and the local human fishers have learned to fish cooperatively.[5] The human fishers tap on the side of their boat to summon the dolphin, which then drives fishes toward the boat. The humans catch the fishes in a cast net and split the catch with the dolphin. They can afford to be generous because they catch

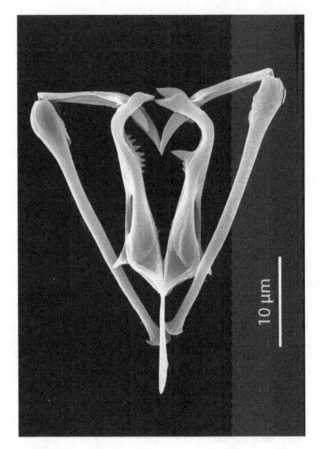

Fig. 11.2. An example of the scary mouthparts of predatory rotifers: the mouthparts of *Dicranophorus capucinus*. *Source:* Willem de Smet.

2–3 times as many fish when working with the dolphins than when working by themselves.

Inland waters contain many smaller and even scarier predators that you probably haven't heard of. A predatory rotifer, for instance, would be a terrible thing to have chasing you. Although less than an eighth of an inch (3 millimeters) long, these animals grab their prey with special tongs (fig. 11.2) that look like they were designed by the inventor in *Edward Scissorhands*. The shapes of these tongs differ across rotifer species, and some have oddly particular shapes, with notches and teeth that must be matched to the special prey of that rotifer.[6] If you were a potential prey item, you'd have to wonder if these distinctive structures were designed *especially for you*.

The flatworms are probably my own worst nightmare among the inland-water predators.[7] Flatworms include the planarians that you may remember from biology class, but there are more than a thousand species of flatworms of various shapes and sizes in inland waters. Based on what you know about planarians, you may think that flatworms are kind of cute. Sure, they're cute—the same way that the clowns in horror films are cute. Flatworms glide silently around sediments and plants searching for prey. Some have those creepy planarian eyes, some have no eyes, and some have dozens of eyes scattered around the margins of the head and body. If they don't see you, they can smell you or feel you as you move around (hold very still, and *don't make an odor*).

When some flatworms encounter their prey, they swallow it whole, or stick out their mouthparts like a vacuum-cleaner hose to slurp up the tender parts of their prey or plunge them into the prey's body and suck out its contents. Some of these species have several of these hoses (yeek). There are species that entangle their prey in sticky or paralyzing mucus (probably my own least favorite death, to lie paralyzed in mucus while a voracious predator approaches), and a group of flatworms called kalyptorhynchs that have a proboscis that is sticky or armed with hooks that they use like a lasso to grab and stuff prey into their waiting mouths. Other flatworms stab and subdue or kill their prey using their sharp penises (I don't even know what to say about that). One common flatworm uses weapons that it steals from another animal to capture its prey. *Microstomum* likes to eat *Hydra*, a freshwater cousin of jellyfishes. Like jellyfishes, *Hydra* has stinging cells called nematocysts that it uses to sting and immobilize its prey—sort of a biological taser. When a *Microstomum* eats a *Hydra*, not all of the nematocysts are fired, and the flatworm saves some of these stinging cells, which it embeds in its own body to use, probably both for defense and for capturing its own prey. Scientists still don't know how *Microstomum* keeps from setting off the nematocysts and getting stung.

And these scary predators are *everywhere*. They live in all kinds of fresh and brackish waters, and often are abundant. In a small New Hampshire lake that I studied, there were 17 flatworms per square *inch* of lake bottom, on average, so there is little hope of avoiding contact with flatworms.[8]

So if you want to make a really scary movie, don't send in the clowns—use people-sized flatworms instead.

Predation is so prevalent in inland waters that analyses by scientists have concluded that the fate of most inland-water animals is to be eaten by a predator, not to die of old age, be swept away by currents, or get crushed by a falling anvil.[9] So the question for many inland-water animals is not whether you will die by being eaten, but just what will eat you and when. If you ever find yourself somehow transported into *Honey, I Shrunk the Kids*, *don't go into the water*.

DIGRESSION 11.1

Inland Waters Where the Cupboard is Bare

Some inland-water environments contain so little food that it can be a challenge to find *anything* to eat. Frigid lakes in rocky basins poor in the nitrogen or phosphorus needed to nourish plant life produce so little food that any fishes unfortunate enough to be living there are likely to be scrawny and slow-growing. But the most food poor of the inland-water habitats are the deep groundwaters.

Any food that reaches deep groundwaters from the sun-fueled ecosystems of Earth's surface must either have been released as sedimentary rocks slowly decay, a process that liberates bits of third-rate food produced by photosynthesis millions of years ago, or percolated through overlying soils, then moved slowly through sand, gravel, or stone, sometimes taking thousands of years to reach the spot where some poor groundwater creature or microbe is trying to eke out an existence. Anything good to eat was consumed a long time ago. Think about what would be left in your grocery store after people had picked through it for a few months—that's what groundwater organisms get to eat.

Although we surface dwellers think of the beneficent rays of the sun as the ultimate source of food and energy for all life, some of the food in groundwater ecosystems comes from darker sources. Geological processes deep in Earth produce substances like hydrogen, methane, and carbon monoxide, which when combined with chemicals from Earth's crust such as sulfate, iron, manganese, and carbon dioxide, can provide the energy and building materials needed to support life.[10] These chemicals may not sound to you like the items on a buffet table, but specialized groundwater microbes are able to use them as food in the same way that we use an apple and the air that we breathe for sustenance. These deep groundwater microbes are among the few forms of life on Earth that do not rely at all on the sun's energy.

But even with the addition of such Hadean offerings to the meager leavings from surface ecosystems that reach deep groundwaters, there is little food of any kind in most groundwaters, and much of it is crummy. As a consequence of such severe food limitation, population densities of microbes and animals in groundwaters tend to be low and drop rapidly as you move into deeper groundwaters.[11] Indeed, many deep groundwaters have no animals at all and support only microbes at population densities thousands to millions of times lower than in surface waters.

Furthermore, the species that do live in groundwaters tend to lead long, slow lives.[12] Bacterial populations may have doubling times of years to centuries rather than hours to days, and animals that live for a year or in surface waters have relatives in groundwaters that stretch out their lives over a decade or more. Another way that groundwater species adapt to the scarcity of food is through small body sizes that require less food. Some groundwater bacteria and archaeans are so small (for fans of small numbers, these bacteria are just 0.1 micrometers long [that's $1/250,000$ of an inch], compared to the gargantuan 2 micrometers [$1/13,000$ of an inch] of the well-known *E. coli*), with tiny genomes packing only the most essential genes, that they are classified by conventional methods as "dissolved" rather than as particles and were discovered only recently.[13]

Although some of the examples I've offered are striking, all of them are more or less familiar ways of feeding—eating other organisms, whether plants, algae, animals, or bacteria. Some inhabitants of inland waters get their lunches in less familiar ways. For instance, a few inland-water animals can photosynthesize like plants, and convert sunlight to food. (Ok, it's not really the animal that is photosynthesizing but rather algal cells that live right in the animal's tissues. The end result is the same, though—the animal is nourished by the products of photosynthesis that occurs inside its body. And if you want to object that this doesn't count as the animal photosynthesizing because symbiotic algae are doing the work, you may have to rethink your own self-image, because your own respiration is carried out by the mito-chondria in your cells, which are derived from symbiotic bacteria.) In fresh waters, there are green hydras, green freshwater mussels, green flatworms, and green sponges, all of which can photosynthesize like plants.[14] They get some of their nutrition from eating like normal an-imals and some from photosynthesis. This seems like a pretty good deal. If you could photosynthesize, you wouldn't have to run out to get a donut for a snack—you could just step outside into the sunshine. And if you saw someone wearing long pants, a long-sleeved shirt, and a hat, you'd know that they were on a diet.

If some animals can photosynthesize like plants, why shouldn't some plants be able to act like animals and catch prey for food? In-deed, carnivorous plants are common in inland waters. Many of the most famous of the meat-eating plants live in boggy or marshy areas (you often have to get your feet wet to see sundews, pitcher plants, and Venus flytraps in nature), so they might reasonably be classified as inland-water plants.

But perhaps the supreme example of an aquatic carnivorous plant is the bladderwort. Bladderworts are freshwater plants with feathery leaves that live in quiet waters and boggy soils around the world. Many bladderwort species live fully submerged in lakes and ponds, with only the aerial flower appearing like a periscope above the water's surface (we return to this odd behavior in chapter 12). In addition to their feathery leaves and aerial flowers, bladderworts are covered with small, transparent bladders that are traps for small animals (fig. 11.3).[15] These bladders have tough, slightly flexible sides and a hinged trap door that forms a tight seal against the main body of the bladder. The plant pumps water out of the bladder, forming a strong vacuum inside. When a small animal touches the trigger near the trap door, the door opens suddenly, sucking in the animal, then quickly closes.

Fig. 11.3. A bit of a bladderwort sprig with two of the bladders that are used to capture animal prey. *Source:* Michal Rubeš, CC BY 3.0 CZ, Wikimedia Commons.

The whole process of opening and closing takes about $1/100$ of a second. The plant resets the trap by pumping out the water, then digests the unlucky animal at its leisure and absorbs the nutrients that it contained. Although bladderworts "eat" mostly zooplankton and small crustaceans and insects, they also can take prey as large as mosquito larvae and small tadpoles and fishes on occasion. The transition to meat-eating allows bladderworts to colonize habitats too poor in nutrients to support much other plant life.

Now for the really odd stuff. Once they find some food, by whatever means, all of the animals I've mentioned so far work more or less like humans when it comes to extracting energy from that food. We all use molecular oxygen (O_2, which is 21% of the atmosphere and dissolves in water) to oxidize the organic carbon in food to carbon dioxide, which produces the energy that we all need to run the machinery of life. This is the same chemical reaction that occurs when logs are burned in a campfire; it's just slower, cooler, and more controlled in us, so you can think of yourself as a cool, controlled, smoldering fire, if that helps.

Some bacteria and archaeans can use entirely different chemical reactions to obtain energy.[16] As a result, they can live in places like sediments, stagnant lakes, and deep groundwaters where there is no oxygen, habitats where we and most animals would quickly suffocate. They can also thrive in places where edible organic carbon is scarce by eating other "food" instead.

Bacteria and archaeans use many alternative "feeding" approaches, and scientists are still discovering new ones. Some kinds of microbes use nitrate, sulfate, or certain forms of iron to take the place of oxygen when it's absent. Others break down organic matter in the absence of oxygen by using fermentation (some fungi do this as well; you know them as the yeasts that produce wine and beer). If edible organic matter is scarce, there are bacteria or archaeans that oxidize ammonia (like in the bottle under your sink), methane (like in your natural gas supply), or sulfide (the rotten egg smell in stinky marsh sediments) instead of organic carbon. If there is an oxidation-reduction reaction that yields energy using chemicals that are readily available somewhere on or near Earth's surface, there is a good chance that there are bacteria or archaeans somewhere using it to make a living.

Not only do these diverse chemical reactions allow bacteria and archaeans to live in habitats where humans and the other species that are smoldering campfires could not survive but they play essential roles in element cycling on our planet. Here are just two examples. They keep organic matter from building up indefinitely in oxygen-free sediments, which would alter the oxygen and carbon dioxide contents of the atmosphere and thereby affect Earth's temperature, the frequency of wildfires, and many other characteristics of our planet. The microbes that use nitrate instead of oxygen remove valuable fertilizer from farm fields (nitrate is a key plant nutrient), while at the same time removing an important pollutant of inland and coastal waters. It is no exaggeration to say that the planet Earth would be unrecognizable and probably uninhabitable for humans if not for these microbes and their unfamiliar ways of eating lunch.

· · · ·

You can see even from this brief survey how varied the food and feeding habits of inland-water species can be, allowing them to survive in the many kinds of inland-water habitats. If nutrients are scarce, carnivorous plants get their nutrients by eating meat. If oxygen or edible organic matter are scarce, microorganisms fuel their fires using other chemicals. If there are phytoplankton and other edible goodies floating in the water, animals use nets, pumps, and other devices to capture the bounty. Down on the sediments, animals large and small feed on rooted plants or browse on ubiquitous biofilms. And everywhere, predators glide, lurk, sting, snatch, and swallow unlucky prey. Lunch is all around, if you just know how to find it.

Chapter 12

CHALLENGES 4
And Then There's Sex

It's the twenty-first century, and we're all cool and sophisticated about sex. We know all about the infinite variations of human sexuality and cannot be shocked. Well, as it turns out, inland-water organisms have been considerably more inventive than humans when it comes to sex. To be sure, some inland-water species have a very Ozzie-and-Harriet lifestyle, where one male and one female bond for a long time.[1] Probably the iconic example of long-term monogamy is an inland-water species—the swan, which usually mates for life (although divorce and cheating occur even in this paragon of faithfulness). In other species, monogamy is enforced by jealous males, which hold the female in an embrace for the entire time that she is sexually receptive or insert a plug after mating to prevent other males from breeding with "their" female. That's monogamy all right, but not quite as romantic as the faithful pair of swans.

It is more common for inland-water animals to be less exclusive in their choice of mates, with both males and females having multiple partners. This can occur in several ways. In broadcast spawners like clams and mussels, the males simply release their sperm into the water. Depending on the species, eggs are fertilized after females release them into the water, or in the female's gill after she filters in the sperm of whomever is spawning at the moment. In such species, Harriet's various kids may have different fathers, and Ozzie may have children by several different females. Whom you have kids with is determined more by the currents and wind direction than by love.

In other species, the males and females more or less choose their multiple partners. It is very common among inland-water fishes for spawning females to be accompanied by two or more males that release sperm as the female drops her eggs, and males may mate with multiple females. This presumably increases the chances that a female's eggs will be fertilized and makes the offspring of both males and females more genetically diverse.

But a more fundamental question about males than "which male(s) should I choose as a mate?" is "males, who needs 'em?" Males are an evolutionary extravagance because they don't make any babies. A species that consists entirely of females produces twice as many offspring per generation as a species that has equal numbers of males and females. Harriet and Harriet would soon outcompete Ozzie and Harriet. Furthermore, the recombination associated with sexual reproduction breaks up favorable combinations of genes that might be preserved by clonal reproduction. So evolutionary biologists have been looking for a long time for a large advantage that males and sex might provide to offset these apparently large advantages that a clonal, all-female species should have.

It has been surprisingly difficult to find this advantage, and evolutionary biologists are still arguing about why sex arose and persists.[2] Here are some possibilities. Sexual reproduction seems to be especially advantageous in changing or variable environments. If environmental conditions never change, then the same genotype that is best today will be best tomorrow, and it will be advantageous just to clone this genotype. But if environments are changing or if your offspring scatter to live somewhere else, then a new genotype may do better than a clone. This is particularly true because other species in the community are constantly evolving better ways to eat you, or parasitize you, or outcompete you for limited resources, so you need to keep changing to keep up with them. Sexual reproduction is a good way to generate new combinations of genes that can be tested against these constantly changing environments.

Sexual reproduction also is efficient in dealing with new mutations. If a harmful mutation appears, it can be eliminated more quickly in a sexually reproducing species than in a clonal one. On the other hand, if multiple beneficial mutations appear, the genetic recombination associated with sexual reproduction can quickly produce favorable combinations of genes. Although these are potentially potent advantages, evolutionary biologists are not convinced that they have the whole story, and they are still looking for better explanations of why sexual reproduction is so widespread.

Despite the difficulty that PhDs have had in identifying the benefits of having males, nature must have its reasons, because males are very common in many kinds of plants and animals. But these benefits must not be ubiquitous, because many inland-water species have more or less dispensed with males, with a lot less Ozzie and a lot more Harriet.

In one very common pattern, males are produced only when conditions become stressful. For most of the year, the population consists entirely of females that give birth to daughters that are genetically identical to their mothers. There is no wasteful production of males, and the population may grow very quickly, if conditions remain favorable. When things start to go bad—the population becomes overcrowded, food becomes scarce, the pond dries up, or the growing

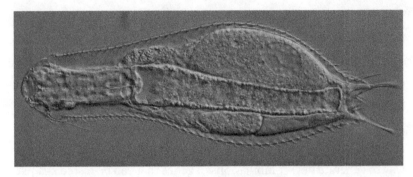

Fig. 12.1. A gastrotrich with developing asexual eggs (the large objects on either side of the gut). Scale bar is 50 µm (about 1/500 of an inch). *Source:* Maria Balsamo, CC BY 4.0, Wikimedia Commons.

season draws to a close, for instance—some of the females produce males that can reproduce sexually with females. This episodic sexual reproduction greatly increases the range of genotypes in the population that will be available to cope with the changing conditions or the unknown circumstances of the next growing season.

The water fleas and rotifers that dominate the zooplankton of lakes are the best-known examples of inland-water animals whose populations are entirely female for most of the year, but many kinds of small invertebrates living on the sediments of lakes and streams have similar life cycles. This episodic sexual reproduction seems like a good way to get some of the advantages of sexual reproduction without paying the full cost of supporting a large population of males year-round. The number of males that are produced may be very small—I worked in a zooplankton lab for three summers handling countless thousands of water fleas and rotifers and never once saw a male.

We don't yet understand all of the interesting variations of this kind of life cycle. Gastrotrichs (fig. 12.1) are tiny invertebrates (most species are less than $1/100$ of an inch [0.25 millimeters long]) that are abundant (often 10 to 100 per square inch [1 to 10 per square centimeter] of lake bottom) nearly everywhere in lakes, marshes, and quiet waters but that are perhaps the animals in inland waters we know the least about. For many years, gastrotrich populations were thought to consist entirely of clonal females (like the bdelloid rotifers I introduce next).[3] Indeed, as far as we know, all inland-water gastrotrichs hatch from eggs as females (when they hatch, they already contain a developing egg, so they are born pregnant). They lay four clonal eggs, genetically identical to the mother, within the first few days of their lives. But then they grow testes and turn into hermaphrodites that make sperm and sexual eggs.

Scientists have no idea how sperm transfer in these gastrotrichs takes place—there are only 32–64 sperm cells in total, no ducts leading outside the body from the testes, and the sperm don't swim, so there must be some special way of transferring the sperm from one animal to another. Gastrotrichs have a big organ of unknown function called the X-organ, which some scientists think may be involved in sperm transfer. (Here's a tip: when scientists name something the "X-organ," you can be pretty sure that they have no idea what it does. They may as well have named it "that big thingamajig.") Another scientist suggested that sperm may be transferred when one animal eats the living body or corpse (ugh) of another. I think that counts as kinky sex in anyone's book.[4]

In other species, females turn into hermaphrodites when males are scarce. Pearl mussels used to cover the bottoms of Bavarian streams, but by the 1980s, they had become hard to find. Gerhard Bauer, who was studying these declining populations, was concerned that the mussel populations might become so sparse that the males and females would be too far apart for the female's eggs to get fertilized. He needn't have worried.

Bauer did an experiment where he took some female mussels and moved them upstream of any males.[5] You might expect that these animals would just stop reproducing, because there would be no way for them to get their eggs fertilized. Instead, Bauer found that many of the females turned into hermaphrodites that produced sperm within 2 or 3 months. In this way, even sparse mussel populations can continue to reproduce.

Mayflies have a different way of dealing with a shortage of males.[6] Most mayfly species have males and females that mate in the normal way (you know, females fly through swarms of males, which have enormously enlarged eyes for detecting the females; think of an especially frenetic singles bar, only with insects), producing fertilized eggs that develop in the normal way. But if the eggs are not fertilized, they develop anyway, mostly into females. (Because of the cellular details of egg development in mayflies, these females are not clones of their mothers, as in the case of zooplankton.) The females resulting from these virgin births can mate with males and produce the usual offspring or continue to produce daughters through virgin birth. So if males are scarce or if the females have troubles finding mates because the weather is terrible during the brief mating season, no problem. Males—take 'em or leave 'em.

Finally, there are a few species that seem to have done away with males altogether, which offends the sensibility of evolutionary biologists to the extent that one scientist called such species "asexual scandals."[7] The most famous example of these asexual scandals are the bdelloid rotifers ("bdelloid" means "leech-like," and refers to the way that these rotifers creep around). Bdelloids are common, microscopic creatures that live among mosses, moist soils, and lake and stream bottoms around the world. Like water bears, bdelloids are *very* resistant to stresses—biologists recently revived some bdelloids that were frozen in Siberian permafrost for 24,000 years.[8]

No one has ever seen a male bdelloid rotifer. Conventional wisdom is that bdelloids have gone for tens of millions of years without males or sex. Bdelloids are scandalous because evolutionary biologists think that sex is necessary to generate genetic variation needed to confront a changing environment and to survive among other species that are constantly evolving. So bdelloids have been getting a lot of attention from evolutionary biologists who are trying to understand how they get along without sex (and to verify that they have indeed gone without sex and haven't been cheating).

Recently, scientists have been examining the genome of bdelloids to look for evidence of ancient asexuality or recent sex.[9] Without going into the details of their analyses, they've found evidence that supports the idea that these animals really have gone without sex for many millions of years, but other evidence strongly suggests that sex *or something like it* (that is, a process that does some of the same things for a species' genetics as sex but isn't conventional sex, which is a stumper) still occurs today. So scientists are still trying to figure out what the heck is going on with the bdelloids (or, to use more formal scientific language: "The mechanism behind the observed signatures of genetic exchanges between bdelloid individuals remains puzzling, and therefore, the significance of outcrossing in this asexual lineage is still unclear").[10]

But if they haven't been having sex, how do the bdelloids maintain enough genetic variation and novelty to evolve and keep up with the demands of living in a changing environment among evolving enemies and competitors? Mutation adds a little genetic variation over the long term, but is too slow to be the answer.

It turns out that bdelloids contain an extraordinarily high amount of foreign DNA—10% or more of their DNA comes from other species—a much higher percentage than in other animals.[11] Sources of this foreign DNA include more than 500 species of bacteria, protists, fungi, plants, and probably other nonbdelloid animals. This DNA was picked up through a process broadly known as "horizontal gene transfer." Horizontal gene transfer between different kinds of bacteria is well known, where it is important in transferring antibiotic resistance across different bacteria, for example. Exactly how it happens in bdelloids and other animals is unknown. The foreign DNA in bdelloids codes for functions like digestion of otherwise indigestible parts of plants and resistance to stress. So bdelloids have unusual abilities that arose not from their own DNA, but from DNA that they've picked up from what you would ordinarily think of as distantly related organisms. Just as you might say that you got your red hair from your grandmother, a bdelloid could say that it got its ability to digest cellulose from its ancestor, the fungus. As important as horizontal gene transfer has been for bdelloids, it does not seem to be a complete substitute for sex, and the mystery of bdelloid sex or chastity remains to be solved.

DIGRESSION 12.1

The Largest Sperm in the World: For What?

Then there are bizarre mysteries related to sex. Ostracods, crustaceans that look like tiny, swimming kidney beans, are widespread and common in inland waters. They have the largest known sperm (compared to their body size) in the animal kingdom, truly enormous sperm that are $1/3$ to 4 times as long as the male's body.[12] (If humans had the same proportions, our sperm would be 2 feet to 24 feet [0.6 to 7 meters] long.) It's tricky to handle such huge sperm, and an extraordinarily large part of the male ostracod's body is devoted to sperm production, storage, and handling. The male ostracod has a large organ (Zenker's organ) specifically for pumping these huge sperm into the female's body. Because the sperm are so large and probably expensive to produce, ostracods make just a few; only a few dozen are transferred to the female during mating. (By comparison, humans typically transfer hundreds of millions

AND THEN THERE'S SEX

of sperm each time they mate.) Furthermore, contrary to the usual image of sperm as little wriggly swimming guys, the huge ostracod sperm are not mobile until they get inside the female's body, perhaps so they don't get tangled and or cause clogs in the elaborate male plumbing.

You'd think that it would be easy for biologists to figure out the adaptive value of such distinctive sperm cells, which must be especially good for something, but biologists remain baffled. Experts recently wrote "thus far, no hypothesis satisfactorily explains the origin of giant sperms in ostracods, or the longevity of this trait through geological eras, and their existence remains enigmatic," which is the long way to say "beats me."[13]

Nor are animals the only inland-water organisms with sexual challenges. Plants have had a hard time solving the problem of underwater pollination. Up here in the air, many plants are pollinated by animals (often bees but also flies, butterflies, bats, birds, and others), which has led to an extravagant proliferation of gaudy flowers and nectar whose purpose is to attract and reward pollinators. As I write this in my August garden, I'm surrounded by the hum of chubby bumble-bees, amber honeybees, exotic-looking wasps, and tiny "sweat bees," a sound punctuated occasionally by the dry rattle of swallowtail and monarch butterflies and the buzz of a hummingbird, all going about their business of pollinating the bright zinnias and tithonias. Earlier in the summer, the air was filled with the pollen of wind-pollinated plants like grasses and pines (as well as the sneezing of hay fever sufferers triggered by all that pollen).

Neither of these approaches works well underwater. There is no underwater equivalent to honeybees. Indeed, until recently, biologists thought that no aquatic plants were pollinated by animals. We now know of an ocean-dwelling seagrass that is pollinated by small invertebrates climbing around in the plants.[14] This is a far cry from the many intricate relationships between plants and their pollinators on land. The absence of specialized pollinators is one of the great differences that separates the underwater world from our familiar world in the air, and it remains a mystery to be solved.

As for the underwater equivalent of wind pollination, the high density and viscosity of water make it much less likely that a grain of pollen simply dropped into the water will reach its target than in the air. So this type of pollination among underwater plants is much rarer than wind pollination among land plants, although a few inland-water plants do use this approach. Instead of flowering underwater, many aquatic plants send their flowers up into the air to be pollinated by insects or wind, just like their land-based ancestors. This looks natural enough for floating-leaved plants like water lilies, where the floating flowers nestle among the leaves. For fully submerged plants, though, the arrangement seems distinctly cumbersome. The flowers are sent up into the air on long stalks, which may be supported by an elaborate system of modified leaves that hold the flower up in the air. The surface of a pond that is home to such species may appear to be populated by disembodied flowers, with no plant in sight. (Honestly, they give me the willies. When I see a pond full of these flowers sticking out of the water, I feel like I'm in the opening scene of a movie called *Watery Grave*, and the director is establishing that this is a creepy place where something bad happened a long time ago.)

But some inland-water plants take advantage of their watery circumstances to aid pollination. My favorite is wild celery.[15] As with many aquatic plants, the female flowers open at the water's surface, on the end of long, flexible stalks. These flowers are water repellant and so produce a tiny dimple in the surface film of the water. The male flowers are produced underwater but then break away from the plant while still unopened and float up to the water's surface, where they open. The male flowers are top heavy, with the pollen-bearing parts at the top. The male flowers sail across the surface of the water, driven by the wind, until they encounter a female. They then tip over into the little depression made by the female flower, and . . . we're going to have babies!

So just as inland-water organism have many ways of avoiding being washed away, managing the drying out of inland waters, and locating food, they also have many ways of engaging in sex. Some of these are clearly matched to the demands of a wide range of habitats and allow inland-water species to thrive across different kinds of situations. However, a surprising number of the most striking sexual structures and practices still are poorly understood (at least by biologists, if not by their users), and more variations on the simple Ozzie-and-Harriet lifestyle are being discovered all of the time. Even the reasons for the existence of sexual reproduction, so common in nature, are not well understood.

• • • •

With these four examples, I end this brief, selective tour of the adaptations of the inland-water biota. We could continue on to look at many other problems faced by species living in inland waters that I have scarcely mentioned. How do you avoid predators? How do you survive in habitats where oxygen is scarce or missing altogether? How do you deal with the poisons that occur in some inland waters? How do you take up essential nutrients when they are scarce? How do parasites and diseases find their hosts, and how do these hosts dodge these enemies? How do you move around? How do you communicate? But I don't think that I need to go through an encyclopedic recitation of inland-water biology to make the points that I am trying to make here; these four examples should do.

No matter which set of adaptations we examined, we would see that inland-water organisms have developed a variety of solutions to even the most difficult problems of survival, which allow different species to persist and thrive across the wide range of circumstances that occur in inland waters. The high numerical diversity of the inland-water biota as described in chapter 8 corresponds to an equally large (and more interesting) diversity of biological traits and adaptations. We would also see that scientists have not yet discovered all of these adaptations and are still trying to solve some of the mysteries of how inland-water life has adapted to conditions on the rest of the blue planet.

Chapter 13

PERIL

Human Impacts on Inland Waters
and Their Biodiversity

I have to begin these final chapters with an apology. So far, I've tried to keep the tone of this book light, with plenty of humor. I'm afraid that this chapter is going to be serious in tone because our final subject is the current state of inland waters and their biodiversity, which is pretty terrible. I just can't make anything funny out of this subject. The main purpose of this book is to convince you that inland waters are amazing, as amazing in their own way as the ocean and the life it contains. If this is all you're interested in, then maybe you should just skip these last chapters.

But I feel that I have to say something about the current poor state of inland waters and their biodiversity, because we are in danger of losing many of the wonderful things that I've been writing about. The choices we make today will determine how badly inland-water ecosystems and their inhabitants are damaged in coming years, so the subjects of inland-water management and conservation are highly relevant to the future of the rest of the blue planet. We can make wise decisions that preserve inland waters and biodiversity for future generations, or we can keep making careless choices that continue the downward course for inland-water species. Facing up to the grim prospects for inland-water biodiversity can help to motivate us to take action to preserve inland waters and the species they support.

So far, our interest in inland waters has been essentially aesthetic— inland waters and their inhabitants are infinitely varied, interesting, and beautiful. This alone should be enough to make us want to admire, study, and protect them. But inland waters are immensely valuable to people in many other ways. They provide drinking water, fish and other foods, gravel and sand for building, irrigation water, highways for transporting people and goods, hydroelectric power, recreation, a place to dispose of our wastes, beautiful settings for our houses and parks, and much more.

The great value of inland waters for so many purposes has meant that they have been used intensively for millennia, serving as the centers of many of the great early civilizations along the Nile, the Indus, the Tigris-Euphrates, and elsewhere, and they still provide the setting (and essential support) for many of the world's largest cities today. This intensive use has strongly affected the character and quality of inland waters, with human use often diminishing the other values of a body of water. The public benefits of inland waters, including their value as habitats for plants and animals, have tended to lose out to uses that generate money in the short term, resulting in large losses of inland-water biodiversity around the world. Several kinds of impacts have been especially harmful to inland-water life.

POLLUTION

Since the dawn of civilization, people have used inland waters as dumping grounds for sewage, industrial wastes, and runoff from agricultural operations. After the Industrial Revolution, water pollution intensified, making many inland waters unsuitable for drinking, recreation, and aquatic life and making it unpleasant even to live near badly polluted waters. During the "Great Stink" of 1858, the Thames River in London was described as "a Stygian pool, reeking with ineffable and intolerable horrors," and Parliament found it difficult to meet nearby, even after curtains on the river side of the building were soaked in lime chloride to suppress the odor.[1] Many other inland waters all around the world were poisoned by a stew of industrial poisons that killed aquatic life, fouled by human sewage or animal waste that stripped the dissolved oxygen from the water as it decomposed, or overly enriched by nutrients, leading to runaway growth of algae, bacteria, and plants. In addition to reducing the value of inland waters as a source for drinking water (who wants to drink from a Stygian pool?), recreation (or swim in one?), and house sites, these poisons and lack of dissolved oxygen eliminated many species from badly polluted waters.

A particularly widespread, harmful, and hard-to-control form of pollution is the excessive loading of nutrients into inland waters, which scientists call "eutrophication." Nutrients, principally phosphorus and nitrogen, come into lakes and rivers from raw or treated sewage, runoff of fertilizers from farm fields and lawns, and rain and snow falling through polluted air. Although these fertilizers occur naturally in inland waters, the amounts arising from human activities far exceed natural levels and cause radical and often harmful changes to ecosystems.[2] Frequently, inland waters receiving high amounts of nutrients are filled with dense growths of phytoplankton, including toxic forms, which can shade out beds of underwater plants. These dense phytoplankton growths can also cause oxygen depletion at night or in the deep waters of a lake, killing fish and other animals.

Recent improvements in industrial processes and waste treatment have reduced the amount of pollution entering inland waters in developed countries, but pollution still damages inland-water ecosystems even there (see digression 13.1). Elsewhere in the world, many inland waters are still badly polluted by sewage, industrial effluents, fertilizers and pesticides from agriculture, mining wastes, runoff from cities, and so on.[3] This pollution strongly affects inland-water biodiversity; for instance, it was recently estimated that about 5% of the length of China's rivers are too badly polluted to support fishes.[4]

Pollution in a Postpollution World

Here in the United States and in many other developed countries, we may be tempted to think of water pollution as an old problem that has been solved. It's true that we have made great progress in reducing water pollution since the bad old days of the mid-twentieth century, as a result of better laws and improved technology, but at least four kinds of pollution are still a problem in inland waters.

First, some of the pollutants that were dumped in the past are so long lived that they still foul inland waters. For example, General Electric dumped something like a million pounds (a half-million kilograms) of PCBs (polychlorinated biphenyls) from its capacitor factories into the Hudson River in the northeastern United States between the 1940s and 1970s, during a time when these releases did not require permits. PCBs accumulate in food chains and have been linked to cancers and neurological and developmental problems.[5] They also last for a long time in nature, so the Hudson was still so badly contaminated by PCBs after the year 2000 that most of its fishes were unsafe to eat. To reduce this contamination, the US Environmental Protection Agency forced General Electric to dredge and remove almost 3 million cubic yards (2.3 million cubic meters) of sediment from the Hudson, at a cost of $1.7 billion. This program reduced, but did not eliminate, PCB contamination in the

Hudson. Such persistent pollution from past activities is common in developed areas around the world and can be expensive or impossible to remove.

Second, pollution that doesn't come out of a single point such as a pipe but instead runs off of farm fields and cities or falls from the sky, which collectively is called nonpoint-source pollution (yet another example showing why scientists shouldn't go into advertising or write poetry) has been much harder to control than point-source pollution. Examples include fertilizer and pesticides from farm fields, salt and oil from roads and parking lots, and acid rain and mercury from the air. Nonpoint-source pollution has caused many problems for inland waters around the world—farm runoff has fueled toxic algal blooms that impair drinking water and kill aquatic animals; salt from roads and other sources has more than doubled salt concentrations in some rivers and has probably begun to affect aquatic life and human health; and acid rain has wiped out fishes and other sensitive animals from many lakes and rivers, to name just a few examples.[6] Nonpoint-source pollution is likely to remain a problem for many decades.[7]

Third, people are always inventing new kinds of pollution, which may evade existing waste control systems.[8] Recent examples that have been in the news include plastics (such as microplastics), which seem to be everywhere, and "emerging pollutants" such as the pharmaceuticals that pass through our bodies and sewage treatment plants and then cause problems in aquatic ecosystems. New pollutants will continue to be a problem as long as we keep introducing new materials into the environment, especially without first exhaustively testing their environmental effects.

Four, shit happens.[9] To be a little more precise, when we store large volumes of hazardous materials and transport them across the landscape, accidents will result in spills into inland waters. Such accidents are common. Here are just a few of many examples. In 1986, a fire at Sandoz's storehouse for agricultural chemicals in Switzerland released tons of pollutants into the Rhine River, including organophosphate insecticides and many other toxic chemicals, along with the firefighting foam that was used to put out the fire.[10] Nearly the entire eel population of the Rhine from Switzerland to The Netherlands was killed, along with many other fishes and wildlife. In 2010, a pipeline break caused more than a million gallons (4 million liters) of heavy crude oil to spill into Michigan's Kalamazoo River.[11] This spill contaminated about 25 miles (40 kilometers) of river and cost more than a billion dollars to clean up.

Spills seem to find biodiversity hotspots the way that tornadoes find trailer parks. The Clinch River in Virginia and Tennessee is one of the few remaining places in the Ohio River basin that still contains much of its original, diverse assemblage of fishes and mussels and is an important refuge for some very rare species. In 1998, a tanker truck overturned along a tributary about 500 feet (150 meters) from the Clinch, spilling 1,350 gallons (5,100 liters) of Octocure-554-revised, a rubber accelerant that turns out to be highly toxic to aquatic animals. The spill turned the river milky white for more than 7 miles (11 kilometers) downstream and killed 18,000 mussels, including 750 animals of three endangered species, as well as many fishes and macroinvertebrates.[12] At the time, this was the largest known kill of endangered species in the United States since the Endangered Species Act had been passed in 1973. Fish Creek runs near the northern end of the Ohio-Indiana border. An endangered species biologist wrote that Fish Creek "may be the most diverse stream community remaining in the Great Lakes watershed" and by the 1990s it was the only place on Earth where the white catspaw mussel still survived.[13] A pipeline rupture in 1993 spilled more than 30,000 gallons (110,000 liters) of #2 diesel fuel into the creek, killing fishes, mussels, birds, reptiles, and amphibians. No living white catspaws have been seen since then.

—

So we've made great progress in curbing pollution in many parts of the world, but even there it is hardly a problem of the past.

DAMS, FRAGMENTATION, AND WATER WITHDRAWALS

Although dams can provide hydroelectric power, irrigation and drinking water, flood control, and other benefits, they have caused enormous damage to inland-water species by blocking movement of fishes and other organisms up and down rivers, converting running-water habitats to reservoirs, allowing water to be diverted from the river to farm fields and cities, and changing flow patterns, temperatures, and sediment transport for miles downstream. Probably no other single cause has led to the extinction of more inland-water species than dams.

The number of dams on the world's rivers is almost unimaginable; there are close to a million dams worldwide, about 60,000 of them behemoths more than 50 feet (15 meters) tall.[14] The largest dams are more than 800 feet (240 meters) tall, and single dams can hold back more than 10 *cubic miles* (*40 cubic kilometers*) of water, flooding river valleys for more than 100 miles (160 kilometers) upstream. Even small dams interfere with movements of stream-dwelling animals, and bigger dams stop essentially all fishes and other swimming and crawling animals from moving upriver. (The fish ladders that you've seen pictures of allow only the most athletic species to move upriver, and don't help baby fishes move back downstream through the reservoir and over the dam, so they are far from adequate in undoing the harm caused by dams.)[15] Migratory animals like salmons, shads, and sturgeons have been especially badly harmed by dams, with many populations disappearing altogether.

The reservoirs behind dams little resemble the river they replace. Typically, water flowing over the river shallows is replaced by still, deep water; clean sand, gravel, and cobble is covered by mud; and dissolved oxygen may drop or disappear, causing most of the original riverine inhabitants to leave or die. When the Tennessee River and the Cumberland River in the southeastern United States were largely converted to a series of reservoirs, leaving little free-flowing habitat remaining, dozens of species of snails and mussels went extinct, and other formerly abundant riverine fishes and shellfishes were able to persist as small populations only in headwaters or tributaries.[16]

The great river rapids and their inhabitants have been hit especially hard. Rapids of the largest rivers are attractive places to build dams, because these are the sites with the highest potential to generate hydroelectricity. Except in recently glaciated regions (Canada and Siberia, for example), the great river rapids often support distinctive life-forms that are found nowhere else. These specialized species are typically lost when their flowing-water home is converted to a silt-bottomed pool. For instance, the Muscle (or "Mussel") Shoals of the Tennessee River, named for its abundant shellfish populations, supported 70–80 species of freshwater mussels, the most anywhere in the world, before the shoals were drowned by a series of dams.[17] The shoals are now under several feet of silt at the bottom of an impoundment, and 9 of the species that used to live there are now globally extinct. Many of the remaining great rapids on rivers like the Mekong and the Congo are slated to be destroyed in the next few decades by dams now being built or in the planning stages.[18]

It's natural to think of fragmentation by dams (and other causes) as a problem solely for migratory species whose movements are blocked. But fragmentation can combine with other problems such as pollution or climate change to create long-term problems for many species. Consider what happens when there is a pollution spill in an unfragmented river—the pollutant kills animals around the spill site, then animals move back in from unaffected reaches elsewhere in the river system. The pollution eliminates the local population only temporarily. But in a fragmented river, animals can't get over dams or other obstacles to get back into the poisoned reach, and the population is permanently lost. The same holds if a local population is lost through drought, predation from an introduced predator, over-harvest, or any other cause. When the climate changes, animal and plant populations may move across the landscape to seek suitable conditions. In a fragmented river system, such movement is difficult or impossible, preventing migration in response to climate change and thereby increasing population losses and extinctions. Fragmentation thus has the effect of vastly multiplying the severity of almost any other threat.

Dams that are operated to generate hydropower, control floods, or allow water to be diverted to irrigation or cities can produce unnatural patterns of flow for many miles downstream (fig. 13.1). A fish living in today's Colorado River could tell you whether it's a weekend or a weekday by how much water is in the river, but not what season it is! In extreme cases, no water at all is released, and long reaches downriver may be dry or barely damp for much of the year, which is hardly conducive to aquatic life.

Some dams release cold water from the depths of the reservoir, which may be good for introduced trout but bad for the native life of the river downstream of the dam.[19] A 150-mile (250-kilometer) reach of the Cumberland River in Kentucky and Tennessee runs so cold that the pearly mussels haven't reproduced since 1952, though these aging (and presumably, shivering) mussels will reproduce if you move them to a warm place.[20]

Dams trap sediments moving down the river, which can cause problems downriver where the ecosystems are starved of sediments.

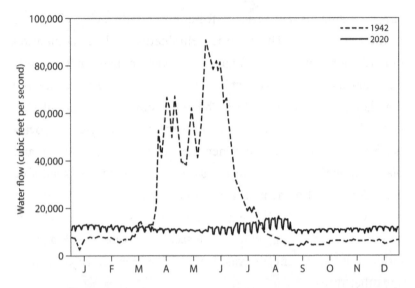

Fig. **13.1.** Dams can greatly affect flow patterns in rivers. This example is from the Colorado River at Lees Ferry, Arizona, before and after dams were built upriver. Before dams (broken line showing data from 1942), the spring snowmelt fed high flows, which cued fish spawning and other biological activities. After dams (solid line showing data from 2020), flows were low and steady year-round, except on weekdays, during which the flows were a bit higher (owing to high electrical demand) than on weekends. *Source:* Graph based on data from the US Geological Survey.

Sediment-starved rivers dig down into the landscape, threatening human infrastructure (bridges, pipelines, etc.) near the channel. Fine sediments are swept away and not replaced, removing habitat for burrowing animals like dragonflies and pearly mussels. Farther downriver, deltas on rivers like the Nile, the Mekong, and the Mississippi that are robbed of sediment by upriver reservoirs sink and erode away.

Often done in conjunction with dams and reservoirs, water withdrawals for irrigation, municipal use, and other purposes can be very harmful to inland waters. In extreme cases, so much water is taken away that the body of water disappears. The most famous example

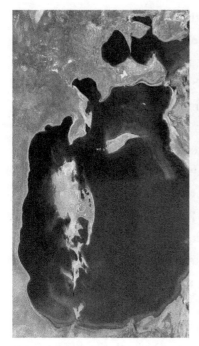

Fig. 13.2. The Aral Sea in 1989, after it had already shrunk substantially (*left*), and in 2014 (*right*), after withdrawals of water for irrigation had nearly drained it out of existence. The length of the sea in the left photo is about 200 miles (320 kilometers). *Source:* NASA; collage by Producercunningham, Wikimedia Commons.

probably is the Aral Sea, once the fourth-largest body of inland water on Earth.[21] The Soviet Union built large (and inefficient) irrigation works, which diverted much of the inflowing water from the lake. As a result, the sea shrank dramatically between about 1960 and 2010 (fig. 13.2), killing much of the native biota and leaving seaside villages high and dry. Efforts since the year 2000 have succeeded in restoring the North Aral Sea, which occupies about 5% of the original area of the Aral Sea. The area once covered by the eastern part of the Aral Sea is now called the Aralkum Desert, which hardly sounds like a suitable home for aquatic species.

Other famous examples include the Colorado, Brahmaputra, Nile, and Yellow Rivers, all of which have stopped flowing to the sea for long periods because of excessive water withdrawals, and shrinking salt lakes such as the Great Salt Lake, the Dead Sea, and Lake Urmia, a salt lake in Iran that was once an important stopover for migratory birds.[22] Countless smaller, less famous wetlands, lakes, and streams have shrunk or disappeared. And out of sight beneath our feet, aquifers are being depleted, eliminating habitat for groundwater organisms and leading to the disappearance of springs and groundwater-fed brooks.[23]

And more dams are on the way. Recently, there has been a lot of publicity about dam removal projects, chiefly in North America and Europe, which have improved habitat and removed barriers to fish passage. Globally, though, dams are still being built faster than they are being taken down, so the number of dams worldwide is still rising.[24] Particularly troubling are the many dozens of new or planned large dams in biologically diverse rivers like the Mekong, the Amazon, and the Congo, which are likely to have catastrophic effects on the unique animals living in these rivers and threaten the welfare of the many people who depend on these rivers.

Dams are not the only things that have physically changed inland-water habitats. Streams have been straightened, hemmed in by levees, or confined into buried pipes; wetlands and shallow lakes have been drained or filled; complex natural shorelines have been replaced by hard walls of steel or concrete; and sand and gravel mining in and around stream channels causes instability in streams channels that can last for decades and stretch for miles up- and downstream. These and other physical changes to habitats have harmed many inland-water species.

OVERHARVEST

Many inland waters are shallow and accessible to people, so people have been taking fishes, waterfowl, plants, and other resources from inland waters for countless millennia. For a long time, such harvests were unregulated, the only rule being "take as much as you can." The easy accessibility of inland waters and the lack of effective regulations eventually led to a "tragedy of the commons" problem in which valuable species were badly overharvested, especially after human populations grew, fishing technology became more effective, and barriers like dams concentrated migrating fishes. The typical effect of overfishing is to reduce the number and average size of fishes, at which point the fishers often work harder (more effort, finer-meshed nets, and generally less selective gear), leading to even more severe impacts on fish populations.[25] Large, slow-growing fishes are especially susceptible to overharvest, to the point that almost all of the largest-bodied inland-water fishes (e.g., sturgeons, huge catfishes) are now declining, imperiled, or extinct.[26] However, in intensively used and poorly regulated fisheries, even small fishes are overharvested. Overharvest has also harmed species other than fishes, including turtles (taken for meat as well as for traditional medicine) and pearly mussels (harvested for their shells to make buttons and freshwater pearls, as well as for their flesh; in 1913, almost 30 million pounds [13 million kilograms] of shells were taken from living mussels in Illinois).[27]

Inland-water fisheries (including shellfisheries) in much of North America, Europe, and Australia are now routinely regulated, and some overfished populations that were not wiped out during the era of overfishing are now recovering. Globally, though, catches of fishes from inland waters are still rising, chiefly as a result of intensifying fisheries in Asia and Africa, and overfishing is still a serious problem for fishes and other inland-water organisms in much of the world.[28]

BIOLOGICAL INVASIONS

Another important way that humans affect inland-water ecosystems is by carelessly moving species around the world, either by deliberately introducing "desirable" species or accidentally introducing nonnative species. Accidental introductions include species that move through canals, attach themselves to boats and recreational equipment, travel in untreated ballast water, are dumped from aquaria and bait buckets, and so on. Inland waters in densely populated regions often contain dozens to hundreds of these nonnative species, including some of the most conspicuous and dominant species. Brown trout, largemouth bass, multiple kinds of carp, zebra mussels, several kinds of crayfishes, water hyacinth, phragmites, many other kinds of harmful aquatic plants, and many species that cause troublesome diseases in people and wildlife (e.g., human schistosomiasis, crayfish plague, whirling disease of fish) are just a few examples of inland-water invaders that have been widely spread around the world. Because controls on species introductions typically are weak (if they exist at all), current invasion rates in most of the world are high and rising, so we can expect impacts of these biological invaders to increase, at least into the near future.[29]

The ecological impacts of invaders are so strong and varied that they are difficult to summarize in just a few words.[30] Different species of invaders affect water clarity, dissolved oxygen and other aspects of water chemistry, currents and water movement, and even the amount of water in the ecosystem. They compete with, eat, or carry diseases into the native species. They change the amount and kind of food that is available in the ecosystem. It is hardly a surprise, then, that they affect the other species living in inland waters, in some cases eliminating one or more of the native species. Probably only habitat change (from dams, pollution, and the like) currently imperils more species than biological invasions.

It's hard to overstate the strong, pervasive effects of some of these invaders. After the zebra mussel appeared in the Hudson River in 1991, my colleagues and I saw the amount of phytoplankton fall by

75%, the amount of fish food in the river fall by half, a population of 1.1 billion native pearly mussels dwindle nearly to nothing, and variables from water chemistry to fish growth rates change dramatically.[31] Although it has been possible to control or eradicate some of the worst invasive plants in inland waters (usually by using herbicides or biological controls; see digression 14.1), other invaders have been very difficult to control, so their impacts are best regarded as long lasting or permanent.

CLIMATE CHANGE

Of course, climate change is going to cause problems for inland-water species, although we are just now beginning to see its effects. We already are seeing widespread signs of warming—higher temperatures in lakes and rivers, shorter periods of ice cover, and so on.[32] Although the amount of warming so far has been modest, we're seeing signs that it is already affecting inland-water biodiversity—cold-water species are disappearing from places where waters are warming, and warm-water species are moving into waters that were once too cold for them. Scientists expect these shifts to increase dramatically as climate change accelerates in the next few decades. Thus, the cold-loving bull trout has already disappeared from 18% of the streams in Montana where it used to live as these streams have warmed and is predicted to disappear from an additional 39% of occupied streams by the year 2080.[33] On the other hand, the cold-sensitive nonnative basket clam (*Corbicula*) has been expanding all along its northern range margin in the United States and Canada and is predicted to be able to colonize large areas of the Upper Midwest, southern Canada, and northern New England as climates further warm.[34] Models suggest that climate warming will ultimately cause very large shifts in the geographic ranges occupied by inland-water species, if the species

can indeed make their way to hospitable habitats across a badly fragmented landscape.[35]

Climate change will have far greater and more complex effects on
inland waters than simply warming them.[36] The amount and timing
of rain- and snowfall and evaporation will change around the globe,
which will affect the amount of water in lakes, streams, and wetlands.
We also expect to see more severe storms and droughts in many regions. In addition to these obvious effects, climate change will have
countless subtle but important effects on inland-water ecosystems—
changes in lake stratification and mixing, riparian vegetation, and the
timing of leaf fall all have the potential to profoundly affect the species
that live in inland waters.

Further, people are going to respond to water shortages, larger
floods, and other effects of climate change in ways that are hard to
predict but that will affect inland-water species. For example, I expect that water withdrawals from reservoirs and other surface- and
groundwaters will increase, and large-scale water diversions, like
China's South-North Water Transfer Project, will spread invasive
species across the landscape.[37] It seems likely that flood defenses
and flood-control projects will harm inland-water ecosystems, especially if they are designed and built in haste following a catastrophe.
Land uses in many regions will have to change in response to climate
change; we will see shifts in the kinds of crops that are grown and how
they are grown, and these land-use changes will affect inland-water
ecosystems.

Whether arising from warmer temperatures, changed hydrology,
or secondary human response to climate change, the effects of climate
change are likely to be especially severe in inland waters because these
habitats are essentially islands that have been further fragmented by
dams and other human-erected barriers.[38] As a result, it will be difficult

or impossible for many inland-water species to migrate on their own to higher elevations or higher latitudes to find the climates that they will need in the new, warmer world.

LAND-USE CHANGE

Finally, because inland waters are so closely connected to their surrounding landscapes, land-use changes have caused large changes to inland waters and the organisms they contain.[39] Changes in land use have been profound, have affected nearly every part of the globe, and are expected to continue or accelerate in coming decades. The effects of replacing forests and grasslands with cities and farm fields include warmer waters, higher inputs of the nutrients that fuel the growth of algae and aquatic plants, increased exposure of aquatic plants and animals to pesticides and poisons running off of the land, changes in water flows in rivers and water levels in lakes and wetlands, and the losses in (or at least changes in) the food coming in from the watershed.

The river where I do most of my fishing (the Maumee River in Ohio) today is a muddy river full of farm fertilizers, whose water levels swing abruptly from so low that you can wade from bank to bank to a roaring river more than 10 feet (3 meters) deep. Looking at today's river, it's hard to believe that the Maumee of the year 1700 was clear, filled with beds of aquatic "grass," low in nutrients, and had steady flows. The radical changes in the Maumee since 1700 are the result of the conversion of nearly its entire watershed from forests, wetlands, and prairies to row-crop farms, which has driven out fishes that need clear water and the cover provided by aquatic plants. The muskellunge, once common in the Maumee, has long since disappeared, as have the harelip sucker (now globally extinct) and several kinds of minnows with big eyes that found their food by sight, which presumably starved when the water became too muddy for them to see. Before land use practices were improved, so much mud washed off of the plowed fields in spring rainstorms that some of the river's fishes suffocated, their gills packed with silt.[40] Although the river has the same name today that it had in 1700 and it (mostly) occupies the same physical space today as it did then, it is a fundamentally different ecosystem from the one in 1700. This kind of stealth substitution of one kind of ecosystem for another has occurred all around the world as a result of land-use change.

All of these classes of impacts—pollution, damming, water withdrawals, biological invasions, climate change, land-use change—are widespread across the planet, so most inland waters have been subjected to at least one of them. In developed parts of the world, most inland waters have been affected by several or all of these impacts. And although there are some bright spots where things are getting better that we discuss in the next chapter—pollution control, dam removal, better fisheries management—globally, inland waters are being squeezed harder and harder by multiple human impacts, many of which are growing exponentially, reaching levels that modern ecosystems have never experienced.[41]

It is unlikely that these pressures will ease any time soon. Many people in today's world lack adequate drinking water, electricity, animal protein, and other items that inland waters can provide, so even without human population growth, the demands on inland waters would grow. However, the human population (which reached 8 billion today, as I write this) is expected to grow by another 3 billion (3 billion was the total human population of the planet in 1960) by the year 2100. Coupled with economic growth that demands greater goods and services per person, this population growth will increase demand for drinking and irrigation water, hydroelectric power (especially in a world that is trying to wean itself from fossil fuels), fish, and other goods supplied by inland waters. The rest of the twenty-first century will be a challenging time for inland waters.

THE STATUS OF INLAND-WATER BIODIVERSITY

These changes to inland waters have caused large problems for their inhabitants, especially for species that have trouble moving out of harm's way or recolonizing lakes and streams after they've recovered from transient harmful impacts (like chemical spills). As a result, many inland-water species are declining, threatened with extinction, or already extinct. Inland-water species often are faring far worse than their better-known terrestrial counterparts.

You can see this from data from North America (fig. 13.3), where we have pretty good information on at least the larger animals. Animals that do not disperse well (pearly mussels, snails, and crayfishes) are doing very poorly indeed. Dozens of species have already gone extinct (the black part of the pies in fig. 13.3), and something like a third to a half of the species are seriously threatened with extinction in the near future (the dark gray parts of the pies, showing G1 and G2 species). Fewer than a quarter of the species in these groups of poorly dispersing animals are considered to be secure from the risk of extinction (the white part of the pies).

The inland-water species that are better at dispersal (fishes and flying insects) are doing a little better, though you'd hardly say that they're doing well. Just 1–2% of these species are thought to have gone extinct (black slices of the pies), but about 20% of species are seriously threatened with extinction in the near future (the dark gray parts of the pie), and far fewer than half of the species are thought to be secure (white). I suppose that these groups would be thought to

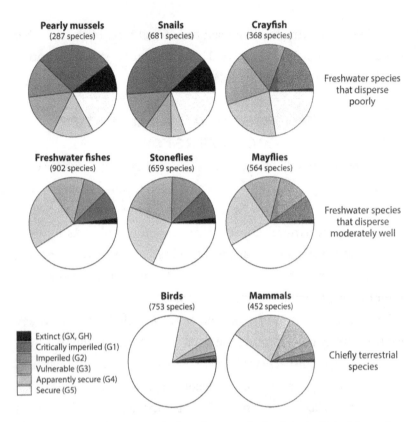

Pearly mussels
(287 species)

Snails
(681 species)

Crayfish
(368 species)

Freshwater species
that disperse
poorly

Freshwater fishes
(902 species)

Stoneflies
(659 species)

Mayflies
(564 species)

Freshwater species
that disperse
moderately well

Birds
(753 species)

Mammals
(452 species)

Extinct (GX, GH)
Critically imperiled (G1)
Imperiled (G2)
Vulnerable (G3)
Apparently secure (G4)
Secure (G5)

Chiefly terrestrial
species

Fig. 13.3. Conservation status of selected groups of animals in the United States (excluding Hawaii) and Canada. Darker shades indicate species that have gone extinct or are in serious trouble. Pies with lots of black and dark gray show groups that are doing poorly; pies with lots of white show groups that are doing better (though still including some species that are doing poorly). NatureServe conservation statuses are GX (extinct, like the ivory-billed woodpecker and passenger pigeon); GH (possibly extinct, like the Eskimo curlew); G1 (critically imperiled species that are at very high risk of extinction or collapse, like the California condor, black-footed ferret, and North Atlantic right whale); G2 (imperiled species at high risk of extinction or collapse, like the long-eared bat and West Indian manatee); G3 (vulnerable species at moderate risk of extinction or collapse, like the spotted owl, piping plover, and polar bear); G4 (apparently secure species at low risk of extinction or collapse but with some reasons for concern, like the American bison, snowy owl, and peregrine falcon); and G5 (secure species at very low risk of extinction or collapse, like the raccoon, bald eagle, and sandhill crane). Excludes about 5% of all species that for various reasons were not given a usual conservation rank. *Source:* Graph based on October 2022 data from www.natureserve.org.

be in dire straits were it not for the much worse status of mollusks and crayfishes.

Finally, for comparison, the bottom panels of fig. 13.3 show the situation for North American birds and mammals. Many people know that some North American bird and mammal species are extinct (passenger pigeon, ivory-billed woodpecker) or facing serious conservation challenges (black-footed ferret, many whale species, spotted owl, whooping crane), and there has rightly been a lot of concern recently about broad declines in bird populations. But their conservation problems, although undoubtedly serious, are far milder than those faced by inland-water species (far less black and dark gray in their pies and far more white). Fewer than 1% of the bird and mammal species in the continental United States and Canada have gone extinct, just 3–7% of species are listed as G1 or G2 (seriously threatened with extinction in the near future), and well over half of species are thought to be secure.

We'd see the same thing if I showed global numbers or numbers from nearly any continental region we could choose; it's just that the North American data are more complete than data from other parts of the world.[42] Inland-water species are facing real problems if we continue business as usual. We don't have great estimates that include all of the small, obscure species of how many inland-water species around the world are seriously threatened with extinction in the next few decades, but the number must be at least in the tens of thousands, including many species that probably will disappear before they are discovered and before we have the least idea of how they make their livings.[43]

As in the case of the broad statistics about inland-water biodiversity discussed in chapter 8, these broad-brush statistics about its decline conceal stories about each species that has disappeared or declined. The baiji, or Chinese river dolphin, was eliminated from

Fig. 13.4. A pickled specimen of the Chinese paddlefish, just one of the many in-land-water species that have gone extinct as a result of human activities. *Source:* Alneth, CC BY-SA 4.0, Wikimedia Commons.

its only home in the Yangtze River through a deadly combination of hunting, entanglement and drowning in fishing nets, collisions with ships, dams, pollution, and habitat destruction.[44] Several other notable species from the Yangtze have probably gone extinct since 1900, including the Chinese paddlefish (fig. 13.4) and the Yangtze sturgeon (which however still lives in captivity).[45] They may soon be joined by the Yangtze River softshell turtle (fig. 8.2, *top left*), perhaps the world's largest inland-water turtle, whose total worldwide population in the wild and in zoos appears to number fewer than 10 animals.[46] Here in North America, 17 of the 28 species of riffleshell mussels (those are the mussels that catch fishes by the nose; see digression 9.1) have gone extinct, and all but one of the remaining species are critically imperiled (G1), chiefly from damming and other habitat changes.[47] In Lake Erie, a fish called the blue pike disappeared (from overfishing?) before we ever understood what it was. Was it a distinct species? A subspecies? Just a colored form of the walleye?[48] This fish was so abundant (between 1914 and 1957, 350 million pounds [160 million kilograms] of blue pike were taken from Lake Erie by the commercial fishery) that people all over Ohio and the Great Lakes region must have known what it tasted like, even if biologists never studied it well enough to know what it was. The last blue pike was seen in 1964.

However you look at it, inland-water species are doing poorly and are facing a dire future. Is there anything that we can do to improve their prospects? Actually, we can do a lot. Here's a hint: it's not business as usual.

Chapter 14

SOLUTIONS

Protecting and Restoring
Inland-Water Ecosystems

There is both good news and bad news about the prospects for saving our inland waters and the life that they contain. Let's start with the bad news so we can focus on the good. Two aspects of the bad news are especially bad. First, as I described in the last chapter in gloomy detail, human demands on inland waters and surrounding ecosystems are huge and growing, and it won't be possible to eliminate these demands or their harmful effects on inland-water ecosystems. We are not going to be able to do without hydroelectricity, irrigation water, flood protection, drinking water, or navigable waterways nor prevent waste disposal into inland waters or harmful uses of the watershed, at least not any time in the near future. And we have been so slow at addressing climate change that inland waters will be affected by intensifying climate change for at least decades to come.

Second, some of the damage we've already done will be difficult or impossible to repair. We will never get back the dozens of extinct species like the baiji and riffleshells. (Despite all of the talk that you may have heard about deextincting species like the passenger pigeon and the woolly mammoth, which if it works will produce only approximate replicas of these species and not literally the extinct species themselves, there is essentially no hope at all of resurrecting the extinct species of inland waters.[1]) The legions of nonnative species that now dominate many inland-water ecosystems will be difficult to remove—it's hard to imagine how we could remove dreissenid mussels, alewife, and round gobies from the North American Great Lakes or the various carp species from the many waters that they have invaded around the world.[2] Pollution by persistent chemicals (metals in lake and river sediments, nutrients accumulated over the decades in heavily fertilized soils surrounding lakes and streams) can be logistically challenging and exorbitantly expensive to remove, and so it is often impractical to fix. Impacts from this past pollution will persist for at least decades. Likewise for large-scale physical changes (enormous dams, drained wetlands, and channelized waterways, for instance), many of which probably are best thought of as permanent (although I mention a few counterexamples in upcoming pages).

This is indeed very bad news, and I'm afraid that we can expect continued degradation and loss of inland waters and declines in inland-water biodiversity, at least over the near term. But the good news is that we don't have to continue along this course if we choose not to. To be specific, because we've done such a poor job managing every kind of our water demands, there are good opportunities to reduce or even reverse every kind of our impacts on inland waters. It would take a book much longer than this one to describe all of these opportunities, but I list a few examples to show their extent and promise.[3]

I've already mentioned that the worst pollution has often been controlled in developed countries, so events like the Great Stink rarely now occur in these regions. Not only has pollution control made it more pleasant (and safer!) to live near, play in, or drink the water but it has also improved habitat for aquatic life (fig. 14.1). However, we could still do a better job controlling the introduction, storage, and transport of environmentally harmful chemicals. There are still many opportunities to control pollution in less developed parts of the world. In addition to benefiting aquatic life, pollution control would benefit the many people who live in and around inland waters in these regions by making it safer to drink or be around the water and increasing the amount of wholesome food that can be harvested from those waters. Nonpoint-source pollution—the pollution that arrives from farm fields, city streets, and the polluted atmosphere—is challenging but far from impossible to control. We can apply fertilizers and pesticides at specific times and in specific places to ensure that these materials stay on the field instead of running off into streams, avoid using unneeded chemicals, add buffer strips or wetlands that trap or destroy harmful chemicals before they can move downstream, and incorporate retention ponds, swales, and wetlands into urban design to do the same for harmful chemicals in urban runoff, to name just a few approaches. We can also ban or restrict the use of the chemicals that are unacceptably harmful to inland waters (as we did by banning DDT beginning in the 1960s).

There are many ways to reduce the harmful effects of dams. The most obvious is to remove the dam.[4] It has become common to remove dams, especially small, antiquated ones, because doing so both provides ecological benefits and removes the hazard that failing dams present to downstream life and property. In cases where dams can't be removed, dam operations can sometimes be adjusted to reduce harmful ecological effects. For instance, some hydroelectric dams were designed to hold back all of the water during times when electricity is not being generated, producing a downstream river channel that

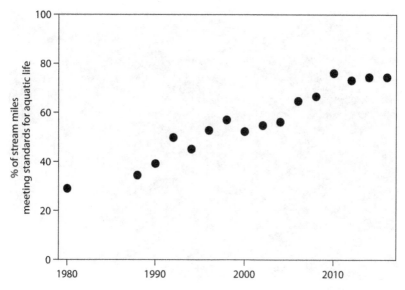

Fig. 14.1. Pollution control improved conditions for aquatic life in Ohio's streams and rivers, allowing many fish species to move back into streams from which they had been driven. Note the improvement from 1980 until about 2010 as the provisions of the Clean Water Act came into effect, then the lack of further improvement as nonpoint-source pollution (chiefly from farms) continued. *Source:* Graph based on data from the Ohio Environmental Protection Agency.

is dry (or at best, damp) for hours or days at a time. Always releasing enough water to keep the downstream channel wetted can improve downstream habitats without substantially reducing the amount of electricity that is generated. Likewise, flow schedules can be adjusted to maximize ecological benefits (providing what are called environmental flows).[5] For instance, extra water can be released at a time of year that promotes fish spawning and rearing. Dams can be designed to minimize harmful changes to downstream water temperatures or to allow some sediment to pass downstream. Further, dams can be placed at sites that minimize ecological damage while still providing the intended benefits of the dam.[6] It is unlikely that dams will ever be

Fig. 14.2. Real rivers have curves. In the 1960s, the Kissimmee River in Florida was straightened into a flood-control canal, which led to ecological problems. To reverse this damage, the canal was filled (the filled canal is the straight, light stripe along the right side of the photo) and the sinuous channel restored. *Source:* US Army Corps of Engineers, VIRIN 091019-A-CE999-001.JPG.

good for aquatic biodiversity, but designing and operating them intelligently, with ecology in mind, can soften their impacts on aquatic life.

We can reduce water withdrawals from inland waters by being more efficient about the ways we use water. There are many ways to be more efficient. Many water supply and irrigation canals are still unlined, which means huge amounts of water are leaked into the thirsty ground. We can line these canals and fix other leaks in water distribution systems. Irrigation systems that lose large amounts of water to evaporation before it ever reaches the crop can be replaced with more efficient drip irrigation. We can meter water and charge the true cost of water to users, encouraging them to be careful how they use water. We know that large gains in efficiency are possible.

For instance, by the 1970s, New York City was running short of water. The city considered building another large water-supply reservoir or drawing water from the Hudson River, but in the end, it chose to use water more efficiently by fixing leaky pipes, installing water meters, and adopting other conservation measures. Water use fell by more than a third (whether expressed as total or per capita water use) between the years 1990 and 2020, freeing up more water than would have been provided by a new reservoir (and avoiding the cost, political difficulties, and environmental damage of building a new reservoir).[7]

Sometimes we can *reverse* the physical damage inflicted on inland-water ecosystems, a very active area of management called ecological restoration.[8] Such restoration may involve restoring the curves to a river that has been straightened (fig. 14.2), exhuming a stream that has been buried in a pipe ("daylighting"), adding rocks or logs to provide habitat for fishes and other creatures, returning water to a wetland that has been drained or filled, replacing a vertical wall with a vegetated shoreline, and so on. Not all ecological restoration has been successful—it turns out to be hard to build or repair ecosystems!—but we can get better at this as we do more of it.[9]

It's always going to be challenging to prevent biological invasions in a globally connected world, but we can reduce the number of new invasions and control at least some of the problematic invaders that are already established in inland waters. With a few exceptions (see digression 14.1), we are doing a poor job controlling the pathways that allow species to move around the world—releases from ballast water, the pet and aquarium trade, poorly secured aquaculture facilities, and well-meaning anglers along with movement on contaminated boats and recreational equipment and along canals. All of these pathways could be better controlled—we can require ballast water to be treated;

regulate the kinds of species that can be sold for pets, aquariums, bait, and aquaculture; educate aquarium owners, anglers, and boaters; set up waterside boat-cleaning stations; impose biological barriers on canals; require more secure aquaculture facilities, and so on. Although many established invaders are difficult to control, sometimes it is possible to control or even eradicate a population if we are willing to spend some money.[10] Here, the island-like nature of inland waters works to our advantage—we may be able to eradicate a harmful species from a single lake without having to eliminate it from an entire continent.

DIGRESSION 14.1

Invasive Species Control in the North American Great Lakes

Biological invasions sometimes are seen as an inevitable result of modern life, but like other human impacts on inland waters, we can control them if we choose to do so. The North American Great Lakes have been heavily invaded by nonnative species: almost 200 nonnative species have been established in the Great Lakes basin, chiefly since the mid-twentieth century.[11] Although most of these species have had small impacts, a few dozen problematic species have had strong ecological and economic effects.

No control has been attempted for most of the established invaders, which are not thought to have troublesome impacts. Other invaders that have had serious harmful effects (common carp and the problematic wetland plant phragmites, for example) have not yet been brought under control, despite repeated efforts. Nevertheless, experience from the Great Lakes shows that biological invasions can be managed or prevented.

Probably the most intensive control effort of an established invader was targeted at the sea lamprey (fig. 14.3).[12] This parasitic fish came

Fig. 14.3. The sea lamprey (two of which are attached to an unfortunate lake trout), a troublesome invader of the Great Lakes whose populations have been reduced through targeted control efforts. *Source:* Great Lakes Fishery Commission, Wikimedia Commons.

into the upper Great Lakes through the Welland Canal. It badly damaged valuable fish populations, nearly eliminating lake trout from Lake Huron and Lake Michigan within a few years of its arrival. Trout landings in these lakes fell from an average of 9.8 million pounds per year (4.4 million kilograms per year) between 1930 and 1945 to just 126,000 pounds per year (57,000 kilograms per year) from 1952 to 1961.[13] Such catastrophic impacts spurred scientists to undertake intensive studies of lamprey biology in the hope that they might find vulnerabilities that could be exploited for control. The insights gained from these studies led to a multifaceted, broad-scale control program that combined barriers and traps on tributary streams with applications of pesticides that were lethal to lampreys but caused relatively little harm to other species. Control measures must be applied every year, are costly, and have some undesirable effects on species other than lampreys, but they have been effective in controlling the invader: sea lamprey populations have fallen by more than 90%, which has allowed valued fish populations to recover.

Purple loosestrife is another species successfully targeted for control.[14] This plant had overrun many wetlands in the northern United States and Canada, crowding out other wetland plants and damaging wildlife habitat. Attempts to control this species through herbicides, water management, mowing, and burning were not very successful, so a biological control program was developed in the 1980s. This program found several species of beetles from purple loosestrife's native range in Europe that ate a lot of loosestrife but did not bother native plants. Following releases of these beetles in the 1990s, loosestrife populations declined in many (but not all) wetlands in the Great Lakes region. The control programs for lamprey and loosestrife show that it is sometimes possible to control troublesome invaders, if you have the right combination of detailed biological research, luck, and tolerance to costs and nontarget effects.

However, it is often less expensive and more effective to prevent invasions than it is to later try to control them. By 1990, it was clear that untreated ballast water was bringing many invaders into the Great Lakes, including some of the most damaging species (zebra and quagga mussels, round gobies).[15] By the late 1990s, the states and provinces around the Great Lakes had agreed to require ships entering the Great Lakes to treat their ballast water. Since then, ballast-water invasions into the Great Lakes have nearly stopped (fig. 14.4). These measures probably have already prevented dozens of new invasions, avoiding the need to later mount targeted control campaigns against any that turned

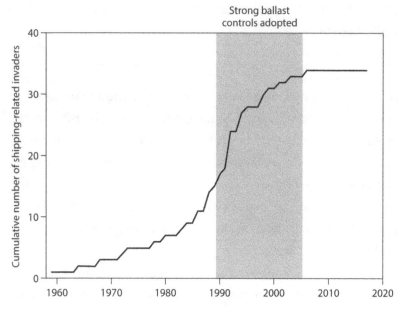

Fig. 14.4. Strong controls on ballast water in the late twentieth century sharply reduced the number of new invasions into the North American Great Lakes and probably have already prevented dozens of new invasions (which you can see by mentally extrapolating the curve from before 1990 into the future). *Sources:* Graph based on data from Sarah A. Bailey, Matthew G. Deneau, Laurent Jean, Chris J. Wiley, Brian Leung, and Hugh J. MacIsaac, "Evaluating Efficiency of an Environmental Policy to Prevent Biological Invasions," *Environmental Science and Technology* 45 (2011): 2554–61, and Rochelle A. Sturtevant, Doran M. Mason, Edward S. Rutherford, Ashley K. Elgin, El Lower, and Felix Martinez, "Recent History of Nonindigenous Species in the Laurentian Great Lakes: An Update to Mills et al., 1993 (25 Years Later)," *Journal of Great Lakes Research* 45, no. 6 (2019): 1011–35.

out to be harmful. The ballast water controls did not require shutting down trade into the Great Lakes or imposing impractical burdens on the shipping industry.

We know how invasive species are moving into inland waters—ballast water, the pet and horticulture trade, contaminated boats and recreational gear, canals, and so on—and we have remedies that could be applied to each pathway that would slow the movement of invaders and prevent future problems. We just don't always have the political will to apply these remedies.

As I'm sure that you've seen in the news, it is within our means to slow climate change. We can become more efficient in using energy, replace fossil fuels with renewables, fix methane leaks in the energy industry, fly less, use more mass transit, eat less meat; by now, this list is pretty familiar. The climate is bound to change over the next few decades as a result of human emissions of carbon dioxide and other gases, and this will affect inland waters, but we can limit the amount of that change by acting now.

As you can see from this brief list of examples, there are plenty of good opportunities to better protect and restore inland-water ecosystems. Many of these actions have benefits beyond conserving biodiversity: removing old dams can eliminate hazards to life and property; controlling pollution can provide safer drinking water; using irrigation water more efficiently can leave more water in river channels for recreation as well as biodiversity; preventing biological invasions can save piles of money; and so on. Furthermore, as our understanding of inland-water ecosystems and experience with their management and restoration expand and as we develop new technologies, these opportunities and their attractiveness should likewise increase.

What is limiting progress is not so much the lack of opportunity as the lack of political will and the unwillingness to spend money. It seems likely that we would pursue more of these opportunities if the public and politicians better appreciated that inland waters and their inhabitants are imperiled and valuable in their own right and are not just resources to be exploited. We need to start thinking more about how actions we engage in affect the special characteristics of inland waters and the unique biodiversity that they contain as well as about the impacts our actions have on their most visible, traditional uses that directly generate dollars (e.g., hydroelectric generation, water for irrigation and drinking, and sport and commercial fisheries). There have even been a few recent attempts to give bodies of water legal rights to exist and flourish, although these attempts have had little practical effect or have been struck down by the courts.[16]

Which leads us finally to a more positive topic—what you personally can do to help protect and restore inland waters and the wonderful species that they support. Certainly, there are personal actions that you can take that can make things better for inland waters. You probably already know most of these. Try to conserve water by using low-flow showers and toilets and fixing leaks. If you live in a dry region, replace your thirsty lawn with xeriscaping. Reuse gray water if it's legal in your community. Don't flush drugs down the toilet or pour hazardous stuff down the drain, because there is a good chance that it will end up in the water. Clean your recreational gear (kayaks, waders, boats, and trailers) before you move it from one body of water to another to prevent spreading invasive species. Don't dump your aquarium, drain your water garden, or otherwise release unwanted pets or plants into the wild. Go easy on using fertilizers and pesticides on your lawn and garden, particularly if you're near water. Eat less meat. Spread the word about how special and vulnerable inland waters are (a good way to do this is to take a child down to your local pond or stream to pursue the ferocious crayfish or the wily frog or just float sticks down the creek).

But much of our impact on inland waters is determined not by our personal actions but by our collective actions, so political engagement can be an effective way for you to help protect inland waters. There are many examples. You can vote for proenvironment candidates at all levels of government (don't neglect the local officials, whose decisions are often important in determining land uses). You can weigh in on local land use and development proposals or proposed changes to local ordinances. You can contact your elected representatives to support adequate funding for natural resource agencies (which tend to be chronically underfunded), support proposed legislation that protects inland waters, and oppose proposed legislation that could damage inland waters. You can submit formal comments on proposed agency actions that affect inland waters and so on.

In fact, there are so many opportunities for political engagement at all the different levels from local to international that it can be hard to keep up with them all. If you're not a politics junkie who likes keeping up with the progress of all the latest bills and behind-the-scenes machinations (and who knows *how* to keep up with such things), a good way to have some influence in the political arena is to join one of the many organizations or advocacy groups that protect inland waters. Many of these organizations weigh in on political matters, and some have government relations staff who keep up with the latest developments in state capitals, Washington, DC, and international forums. They also typically do a lot of good work outside of the political arena, such as buying and managing nature preserves and running educational programs.

I'm going to get in trouble by giving some examples of these organizations, because I'll inevitably leave out somebody's favorite group (besides, the list of the most active players changes over time), but here are a few to consider. At the local level, look for watershed protection groups and lake associations organized to protect your favorite body of water, your local Riverkeeper, or local chapters of organizations that protect aquatic resources (e.g., Trout Unlimited). At the national and international level, check out the websites of the big nature-protection groups (e.g., the Nature Conservancy, World Wide Fund for Nature, Sierra Club, Conservation International, Center for Biological Diversity, and others) and see if they have a program devoted to the protection of inland waters. Finally, there are some national and international organizations devoted especially to the protection and wise management of inland waters (e.g., American Rivers). These groups (and many others that I haven't mentioned, sorry!) do much good work in protecting inland waters and can help ensure that your voice and desires are heard in the halls of political power. If you join one of these organizations, be sure to tell them that you're especially concerned with protecting inland waters.

But whatever your choice, do get involved in some way if you care about inland waters. Otherwise, our society will never give these ecosystems the attention and protection that they deserve.

Chapter 15

BACK TO THE THEME
Closing Remarks

Here, I reluctantly end our brief tour of the world's inland waters. We haven't had the time to explore killer lakes, drumming stoneflies, archerfish, and probably a thousand other fascinating aspects of inland waters, but I've already taken enough of your time. I hope that you've seen enough to convince you that the tiny blue bits of our planet that are the inland waters are astonishingly diverse: volcanic lakes more corrosive than battery acid; rivers torn apart by violent floods then shrunk to their dusty beds by droughts; ponds whose lives are measured in days and rivers that have endured for millions of years; great river rapids, vast, stagnant swamps, and mysterious groundwaters miles beneath our feet—all of them filled with an array of species adapted to this broad range of environments. This array of species includes plants whose leaves are large enough to support people, "aquatic" species that can survive for months, years, or even centuries without water (including tiny creatures so tenacious of life that they can survive a trip to outer space), microbes that survive without the benefits of sunlight, plants that eat animals, a startling array of sexual practices, thousands of catfishes of all sizes, and many equally interesting species that I haven't had the time to describe. Furthermore, we know that there is much still to be learned about inland waters and their inhabitants—many thousands of species have yet to be discovered, inaccessible habitats like the great river rapids and deep groundwaters remain to be explored, and every year, scientists make surprising new discoveries about the adaptations and functioning of even "well-known" habitats and species. This whole array of habitats and species is imperiled by human activities. Inland waters encompass far more diversity, have far more significance for biodiversity on Earth, and still hold far more secrets than one might expect from their modest dimensions.

No single lake or stream is as marvelous as the ocean. But there is only one ocean, and there are millions of inland waters, which taken together contain marvels comparable to those of the ocean. To return to the musical analogy of chapter 1, we could think of the ocean as a magnificent symphony being played in a grand concert hall by a full orchestra. Inland waters are the countless smaller ensembles playing in more modest settings: string quartets and piano trios playing variations on the themes of that symphony in small concert halls; jazz quartets riffing in cafes; choirs in little country churches singing "Joyful, Joyful, We Adore Thee"; even a child playing a simple version of one of the symphonic melodies on a pennywhistle in her backyard. Nothing as grand as the original, perhaps, but taken together an extraordinary variety of pieces and musicians nevertheless. (Or, less pretentiously, you might say that we have fat creeks, skinny creeks, creeks that run on rocks; rough creeks, fishy creeks, even creeks that teem with crocs.)

We may fail to appreciate the importance of inland waters partly because they are such a small part of our planet but also partly because we are so comfortable and familiar with these modest ecosystems (fig. 15.1). Most inland waters aren't majestic, remote ecosystems that we

Fig. 15.1. A modest part of the blue planet (Webatuck Creek, New York). *Source:* Dave Strayer.

see only in nature specials on TV or travel to visit on a dream vacation; rather, they're just the creek that runs behind the Sunoco station, the pond in the woods where we collect tadpoles, the lake where we catch bluegills and teach our kids to swim, the muddy river that floods the corn fields in the spring. It's hard to imagine that there is anything really special or unique about these streams or lakes that we think we know so well.

A few years ago, I drove down to Virginia to take some samples from the Clinch River, one of the most biodiverse remaining bodies of water in North America, world-famous among biologists. When I got down there, I pulled over at the first chance I got to have a look at this wonderful river, only to find that people had dumped an old washing machine and plastic garbage bags over the bank into the river, just as they do everywhere. I doubt that the people who dumped this trash had any idea that the Clinch was like a little bit of the Great Barrier Reef, right there in rural western Virginia.

Yet as I hope I've shown you, these familiar, modest ecosystems contain the most remarkable array of organisms. (For instance, when my colleagues and I visited the little creek shown in fig. 15.1, we found a small population of an endangered pearly mussel, not seen in this region since 1845 and just barely hanging on.) Just as the ocean and its inhabitants are worth saving, our inland waters are worth protecting. They provide many obvious, direct benefits to us, in the form of water to drink and irrigate our farms, fish and other foods for our tables, hydroelectric power in a world struggling to free itself from dependence on fossil fuels, highways on which to transport goods, and places for swimming, boating, fishing, and contemplation. More than this, though, Earth's inland waters contain a remarkable and still incompletely explored array of organisms whose ways of life and ultimate value to humans we are still working to understand.

FURTHER READING

If you would like to learn more about inland waters, I recommend the following books as good places to start (in addition to looking at the sources cited in the notes for each chapter). Jerry Closs, Barbara Downes, and Andrew Boulton, *Freshwater Ecology: A Scientific Introduction* (Malden, MA: Blackwell Publishing, 2004), Jacob Kalff and John A. Downing, *Limnology: Inland Water Ecosystems*, 2nd ed. (Duluth, MN: Bibliogenica, 2016), Christer Brönmark and Lars-Anders Hansson, *The Biology of Lakes and Ponds*, 3rd ed. (Oxford: Oxford University Press, 2017), Brian Moss, *Ecology of Freshwaters: Earth's Bloodstream*, 5th ed. (Chichester, UK: Wiley, 2018), Walter K. Dodds and Matt Whiles,

Freshwater Ecology: Concepts and Environmental Applications of Limnology, 3rd ed. (London: Academic Press, 2019), David Allan, Maria M. Castillo, and Krista A. Capps, *Stream Ecology: Structure and Function of Running Waters,* 3rd ed. (Cham: Springer, 2021), Alan Hildrew and Paul Giller, *The Biology and Ecology of Streams and Rivers,* 2nd ed. (New York: Oxford University Press, 2023), and Ian D. Jones and John P. Smol eds., *Wetzel's Limnology: Lake and River Ecosystems,* 4th ed. (London: Academic Press, 2024) are good, college-level textbooks. To get the most from these books, you may need an undergraduate-level background in biology, chemistry, physics, and mathematics, as well as some patience and determination. Florian Malard, Christian Griebler, and Sylvie Rétaux, eds., *Groundwater Ecology and Evolution* (London: Academic Press, 2023) and William J. Mitsch, James G. Gosselink, Christopher J. Anderson, and M. Siobhan Fennessy, *Wetlands,* 6th ed. (Hoboken: Wiley, 2023) provide good college-level introductions to groundwater ecology and wetland ecology, respectively. If you're interested in something shorter and easier than a textbook, you might look at Colbert E. Cushing and J. David Allan, *Streams: Their Ecology and Life* (San Diego: Academic Press, 2001), Nick Middleton, *Rivers: A Very Short Introduction* (Oxford: Oxford University Press, 2012), Brian Moss, *Ponds and Small Lakes: Microorganisms and Freshwater Ecology* (Exeter, UK: Pelagic Publishing, 2017), or Warwick F. Vincent, *Lakes: A Very Short Introduction* (Oxford: Oxford University Press, 2018). If you'd like to learn more about the daunting problems of freshwater biodiversity conservation and its possible solutions, David Dudgeon, *Freshwater Biodiversity: Status, Threats, and Conservation* (Cambridge: Cambridge University Press, 2020), provides a very fine overview.

NOTES

CHAPTER 1

1. If you don't know what I mean by the blue marble photograph, you can see it here: https://commons.wikimedia.org/wiki/File:Blue_Marble_Western_Hemi sphere.jpg.

2. "Biological diversity," often shortened to "biodiversity," refers to diversity at all levels of biological organization (i.e., the genetic or trait diversity within a species within and across sites, across the species that occur within and across sites, of higher-level evolutionary units such as families or classes, or diversity of ecosystem types). In practice, it often is used simply to refer to the number of species in a given area.

3. Christian Lévêque, Thierry Oberdorff, Didier Pauget, Melanie L. J. Stiassny, and Pascal A. Tedesco, "Global Diversity of Fish (Pisces) in Freshwater," in *Freshwater Animal Diversity Assessment*, ed. Estelle V. Balian, Christian Lévêque, Hendrik H. Segers, and Koen Martens (Dordrecht: Springer, 1998), 545–67; Estelle V. Balian, Hendrik H. Segers, Koen Martens, and Christian Lévêque, "The Freshwater Animal Diversity Assessment: An Overview of the Results," in *Freshwater Animal Diversity Assessment*, 627–37; Greta Carrete Vega and John J. Wiens, "Why Are There So Few Fish in the Sea?," *Proceedings of the Royal Society B* 279 (2012): 2323–29.

4. The statistics here come from US Geological Survey, "How Much Water is There on Earth?," https://www.usgs.gov/special-topics/water-science-school /science/how-much-water-there-earth, which in turn is based on Igor Shik-lomanov, "World Fresh Water Resources," in *Water in Crisis: A Guide to the*

World's Fresh Water Resources, ed. Peter Gleick (New York: Oxford University Press, 1993), 13–24.

5.   Statistics on the number and length of lakes and rivers come from John A. Downing, Yves T. Prairie, Jonathan J. Cole, Carlos M. Duarte, Lars J. Tranvik, Robert G. Striegl, William H. McDowell et al., "The Global Abundance and Size Distribution of Lakes, Ponds, and Impoundments," *Limnology and Oceanography* 51, no. 5 (2006): 2388–97, John A. Downing, Jonathan J. Cole, Carlos M. Duarte, Jack J. Middelburg, John M. Melack, Yves T. Prairie, Pirkko Kortelainen et al., "Global Abundance and Size Distribution of Streams and Rivers," *Inland Waters* 2, no. 4 (2012): 229–36, and Peter A. Raymond, Jens Hartmann, Ronny Lauerwald, Sebastian Sobek, Cory McDonald, Mark Hoover, David Butman et al., "Global Carbon Dioxide Emissions from Inland Waters," *Nature* 503 (2013): 355–59.

CHAPTER 2

1.   Some of you may be wondering whether there is an official distinction between a lake and a pond. There is a long history of scientists trying to formally define lakes and ponds based on surface area, depth, presence of rooted vegetation, etc. (for a recent attempt, see David C. Richardson, Meredith A. Holgerson, Matthew J. Farragher, Kathryn K. Hoffman, Katelyn B. S. King, María B. Alfonso, Mikkel R. Andersen et al., "A Functional Definition to Distinguish Ponds from Lakes and Wetlands," *Scientific Reports* 12 [2022], article 10472). None of these approaches is perfect. One popular definition of a pond is that it is a lake shallow enough to have rooted vegetation growing everywhere, which would make Lake Okeechobee a pond, even though it is the second-largest freshwater lake entirely in the continental United States. Rather than adopting a formal definition, I'll just use "pond" to mean a small, shallow lake. I take the same approach to rivers and streams, treating a river as a large stream.

2.   The *Edmund Fitzgerald* was an ore freighter that sank with all 29 of its crew members in a storm on Lake Superior in 1975. It was made famous by the popular song "The Wreck of the *Edmund Fitzgerald*," by Gordon Lightfoot. Curiously, this was the second time that a ship with that name sank in the Great Lakes. The *Edmond* [sic] *Fitzgerald*, a schooner, went down with its crew of seven in Lake Erie in 1883; see https://greatlakespeopleandplaces.com/2017/09/22/how-the-fitzgerald-sank-twice for more details.

3.   The actual estimate is 277 million ponds and lakes larger than a quarter acre (0.1 hectares); see John A. Downing, Yves T. Prairie, Jonathan J. Cole, Carlos

M. Duarte, Lars J. Tranvik, Robert G. Striegl, William H. McDowell et al., "The Global Abundance and Size Distribution of Lakes, Ponds, and Impoundments," *Limnology and Oceanography* 51, no. 5 (2006): 2388–97. Downing et al.'s estimate may be a little too high—a more recent estimate put the number at 117 million lakes and ponds larger than half an acre (0.2 hectares): Charles Verpoorter, Tiit Kutser, David A. Seekell, and Lars J. Tranvik, "A Global Inventory of Lakes Based on High-Resolution Satellite Imagery," *Geophysical Research Letters* 41, no. 18 (2014): 6396–402. But it's still hundreds of millions.

4. Brennan A. Ferguson, Tina A. Dreisbach, Catherine G. Parks, Gregory M. Filip, and Craig L. Schmitt, "Coarse-Scale Population Structure of Pathogenic *Armillaria* Species in a Mixed-Conifer Forest in the Blue Mountains of Northeast Oregon," *Canadian Journal of Forest Research* 33, no. 4 (2003): 612–23.

5. Carl R. Woese and George E. Fox, "Phylogenetic Structure of the Prokaryotic Domain: The Primary Kingdoms," *Proceedings of the National Academy of Sciences* 74, no. 11 (1977): 5088–90.

6. David L. Valentine, "Adaptations to Energy Stress Dictate the Ecology and Evolution of the Archaea," *Nature Reviews Microbiology* 5, no. 4 (2007): 316–23.

7. Recent reviews have recognized 16–39 phylum-level groups in the Archaea, but the number is still changing (mostly growing). For example, see Christian Rinke, Maria Chuvochina, Aaron J. Mussig, Pierre-Alain Chaumeil, Adrián A. Davín, David W. Waite, William B. Whitman et al., "A Standardized Archaeal Taxonomy for the Genome Taxonomy Database," *Nature Microbiology* 6, no. 7 (2021): 946–59, Nahui O. Medina-Chávez and Michael Travisano, "Archaeal Communities: The Microbial Phylogenomic Frontier," *Frontiers in Genetics* 12: (2021): fgene.2021.693193, and Zhichao Zhou, Yang Liu, Karthik Anantharaman, and Meng Li, "The Expanding Asgard Archaea Invoke Novel Insights into Tree of Life and Eukaryogenesis," *MLife* 1, no. 4 (2022): 374–81.

8. Robert E. Schmidt and Erik Kiviat, "State Records and Habitat of Clam Shrimp, *Caenestheriella gynecia* (Crustacea: Conchostraca), in New York and New Jersey," *Canadian Field-Naturalist* 121, no. 2 (2007): 128–32; Robert E. Schmidt, Erik Kiviat, Norm Trigoboff, and John Vanek, "New Records of Clam Shrimp (Laevicaudata, Spinicaudata) from New York," *Northeastern Naturalist* 25, no. 2 (2018): N7–N10.

9. Crveno Jezero—Red Lake. https://www.wondermondo.com/crveno-jezero-red-lake.

10. Constantinos Taliotis, Morgan Bazilian, Manuel Welsch, Dolf Gielen, and Mark Howells, "Grand Inga to Power Africa: Hydropower Development

Scenarios to 2035," *Energy Strategy Reviews* 4 (2014): 1–10. But there is serious discussion about whether harnessing this potential would be a good idea, either economically or ecologically—see Angelo Carlino, Matthias Wildemeersch, Celray James Chawanda, Matteo Giuliani, Sebastian Sterl, Wim Thiery, and Ann van Griensven et al., "Declining Cost of Renewables and Climate Change Curb the Need for African Hydropower Expansion," *Science* 381, no. 6658 (2023): eadf5848, and Jeroen Warner, Sarunas Jomantas, Eliot Jones, Md. Sazzad Ansari, and Lotje de Vries, "The Fantasy of the Grand Inga Hydroelectric Project on the River Congo," *Water* 11, no. 3 (2019): 407, and references cited therein.

11.   Luna B. Leopold, M. Gordon Wolman, and John P. Miller, *Fluvial Processes in Geomorphology* (San Francisco: W. H. Freeman, 1964).

12.   I am referring here to the movie *On Golden Pond*, the popular song "Suzanne" by Leonard Cohen, William Butler Yeats's poem "The Lake Isle of Innisfree," Mark Twain's novel *Adventures of Huckleberry Finn*, *Great Expectations* by Charles Dickens, and the children's book *Miss Nelson is Missing!* by Harry Allard and James Marshall. I could have cited dozens of other examples from popular culture.

13.   Nick C. Davidson, "How Much Wetland Has the World Lost? Long-Term and Recent Trends in Global Wetland Area," *Marine and Freshwater Research* 65, no. 10 (2014): 934–41.

14.   William J. Mitsch and James G. Gosselink, *Wetlands*, 2nd ed. (New York: Van Nostrand Reinhold, 1993).

15.   Martin R. Kaatz, "The Great Black Swamp: A Study in Historical Geography," *Annals of the Association of American Geographers* 45, no. 1 (1955): 1–35.

16.   Soumya Karlamangla and Shawn Hubler, "Tulare Lake Was Drained Off the Map: Nature Would Like a Word," *New York Times*, April 2, 2023.

17.   "Sweetwater Wetlands Park," https://www.gainesvillefl.gov/Parks/Sweetwater-Wetlands-Park. Nitrate is a form of nitrogen that acts as a plant fertilizer and so can lead to "dead zones" and other overenrichment problems in inland waters and coastal marine waters. Wetlands eliminate nitrate when plants take it up and convert it into organic nitrogen-containing compounds (like proteins) or when bacteria in the wetlands destroy nitrate by converting it into nitrogen gas (the harmless gas that constitutes 78% of Earth's atmosphere).

18.   "Sweetwater Wetlands Park," https://www.gainesvillefl.gov/Parks/Sweetwater-Wetlands-Park.

19.   The Convention on Wetlands (Ramsar), "Wetland Ecosystem Services—An Introduction," https://www.ramsar.org/sites/default/files/documents/library/services_00_e.pdf, 2009.

      Any estimate of the global economic value of wetlands should be regarded as very approximate. The $3 trillion/year figure comes from a famous study of the value of Earth's ecosystems: Robert Costanza, Ralph d'Arge, Rudolf de Groot, Stephen Farber, Monica Grasso, Bruce Hannon, Karin Limburg et al., "The Value of the World's Ecosystem Services and Natural Capital," *Nature* 387 (1997): 253–60. If that seems unbelievably high, a more recent study (Nick C. Davidson, Anne A. van Dam, Colin Finlayson, and Robert McInnes, "Worth of Wetlands: Revised Global Monetary Values of Coastal and Inland Wetland Ecosystem Services," *Marine and Freshwater Research* 70, no. 8 [2019]: 1189–94) came up with a minimum estimate of $27 trillion/year. Whatever the exact value, it almost inconceivably large.

20.   US Geological Survey, "How Much Water is There on Earth?," https://www.usgs.gov/special-topics/water-science-school/science/how-much-water-there-earth, based on Igor Shiklomanov, "World Fresh Water Resources," in *Water in Crisis: A Guide to the World's Fresh Water Resources*, edited by Peter Gleick (New York: Oxford University Press, 1993), 13–24.

21.   Andrei P. Kapitsa, Jeff K. Ridley, Gordon de Quetteville Robin, Martin J. Siegert, and Igor A. Zotikov, "A Large Deep Freshwater Lake Beneath the Ice of Central East Antarctica," *Nature* 381 (1996): 684–686.

22.   R. Allen Freeze and John A. Cherry, *Groundwater* (Englewood Cliffs, NJ: Prentice-Hall, 1979).

23.   Joel Podgorski and Michael Berg, "Global Threat of Arsenic in Groundwater," *Science* 368, no. 6493 (2020): 845–50.

24.   "Dow Chemical Company," *Wikipedia*, https://en.wikipedia.org/w/index.php?title=Dow_Chemical_Company&oldid=1170846892, accessed September 18, 2023.

25.   Such highly localized populations of specialized groundwater microbes appear not to have been found (yet)—Lucas Fillinger, Christian Griebler, Jennifer Hellal, Catherine Joulian, and Louise Weaver, "Microbial Diversity and Processes in Groundwater," in *Groundwater Ecology and Evolution*, ed. Florian Malard, Christian Griebler, and Sylvie Rétaux (London: Academic Press, 2023), 211–40—but it is hard to believe that they do not exist.

26.   Thomas L. Kieft, "Microbiology of the Deep Continental Biosphere," in *Their World: A Diversity of Microbial Environments*, ed. Christon J. Hurst (Chaim: Springer International Publishing, 2016), 225–49.

27. Lazare Botosaneanu, ed., *Stygofauna Mundi: A Faunistic, Distributional, and Ecological Synthesis of the World Fauna Inhabiting Subterranean Waters* (*Including the Marine Interstitial*) (Leiden: E. J. Brill, 1985).

28. Christer Erséus, "*Parvidrilus strayeri*, a New Genus and Species, an Enigmatic Interstitial Clitellate from Underground Waters in Alabama," *Proceedings of the Biological Society of Washington* 112 (1999): 327–37; Rüdiger M. Schmelz, Christer Erséus, Patrick Martin, Ton van Haaren, and Tarmo Timm, "A Proposed Order-Level Classification in Oligochaeta (Annelida, Clitellata)," *Zootaxa* 5040, no. 4 (2021): 589–91.

29. Li-Hung Lin, Pei-Ling Wang, Douglas Rumble, Johanna Lippmann-Pipke, Erik Boice, Lisa M Pratt, and Barbara Sherwood Lollar et al., "Long-Term Sustainability of a High-Energy, Low-Diversity Crustal Biome," *Science* 314, no. 5798 (2006): 479–82; Dylan Chivian, Eoin L. Brodie, Eric J. Alm, David E. Culley, Paramvir S. Dehal, Todd Z. DeSantis, Thomas M. Gihring et al., "Environmental Genomics Reveals a Single-Species Ecosystem Deep within Earth," *Science* 322, no. 5899 (2008): 275–78; Olga V. Karnachuk, Yulia A. Frank, Anastasia P. Lukina, Vitaly V. Kadnikov, Alexey V. Beletsky, Andrey V. Mardanov, and Nikolai V. Ravin, "Domestication of Previously Uncultivated *Candidatus Desulforudis audaxviator* from a Deep Aquifer in Siberia Sheds Light on its Physiology and Evolution," *ISME Journal* 13, no. 8 (2019): 1947–59.

30. Kieft, "Microbiology of the Deep Continental Biosphere," 236; Pierre Marmonier, Diana Maria Paola Galassi, Kathryn Korbel, Murray Close, Thibault Datry, and Clemens Karwautz, "Groundwater Biodiversity and Constraints to Biological Distribution," in *Groundwater Ecology and Evolution*, 113–40.

31. Yinon M. Bar-On, Rob Phillips, and Ron Milo, "The Biomass Distribution on Earth," *Proceedings of the National Academy of Sciences* 115, no. 25 (2018): 6506–11. For an alternative estimate and a critical discussion about problems in estimating how much life is hidden deep beneath our feet, see Cara Magnabosco, Li-Hung Lin, Hailiang Dong, Malin Bomberg, William C. Ghiorse, Helga Stan-Lotter, Karsten Pedersen et al., "The Biomass and Biodiversity of the Continental Subsurface," *Nature Geoscience* 11, no. 10 (2018): 707–17.

32. Magnabosco et al., "The Biomass and Biodiversity of the Continental Subsurface," 710.

CHAPTER 3

1. G. Evelyn Hutchinson, *A Treatise on Limnology*, 4 vols. (New York: Wiley, 1957–93). Chapter 1 ("The Origin of Lake Basins") of volume 1 is a 163-page account of how lakes are formed.

2.   David G. McCullough, *The Johnstown Flood: The Incredible Story Behind One of the Most Devastating Disasters America Has Ever Known* (New York: Simon and Schuster, 1968).

3.   The account of the Indus River flood follows Kenneth Mason, "Indus River Floods and Shyok Glaciers," *Himalayan Journal* 1 (1929): 10–29, available at https://www.himalayanclub.org/hj/1/3/indus-floods-and-shyok-glaciers/.

4.   Hutchinson, *A Treatise on Limnology*; Richard F. Flint, *Glacial and Quaternary Geology* (New York: Wiley, 1971).

5.   Canada alone contains 879,800 lakes larger than 25 acres (10 hectares), which is more than half of the world's natural lakes of this size; See Mathis Messager, Bernhard Lehner, Günther Grill, Irena Nedeva, and Oliver Schmitt, "Estimating the Age and Volume of Water Stored in Global Lakes Using a Geo-Statistical Approach," *Nature Communications* 7 (2016): 13603.

6.   Lake Alachua's colorful history is described by Lars Andersen in *Paynes Prairie: The Great Savannah: A History and Guide* (Sarasota, FL: Pineapple Press, 2014); for an abbreviated version, see Hutchinson, *A Treatise on Limnology*, 103–4.

7.   For instance, the US Environmental Protection Agency has estimated that about half of the lakes in the continental United States were made by humans, with natural lakes predominating in areas that were covered by glaciers or underlain by soluble rocks, and artificial lakes predominating elsewhere ("National Highlight—Comparing Natural Lakes and Manmade Reservoirs," US Environmental Protection Agency, https://www.epa.gov/national-aquatic-resource-surveys/national-highlight-comparing-natural-lakes-and-manmade-reservoirs, accessed October 3, 2023).

8.   See Hutchinson, *A Treatise on Limnology*, for a discussion of some of these oddball lake origins.

9.   I have not found any quantitative estimates of the number of lakes on Earth throughout geologic time, but it is clear that this number varied greatly as various lake-forming forces waxed and waned. For instance, tectonic forces produced large lakes during the Devonian, the Permian, the Triassic, and other times; see Elizabeth Gierlowski-Kordesch and Kerry Kelts, introduction to *Global Geological Record of Lake Basins*, vol. 1, edited by Elizabeth Gierlowski-Kordesch and Kerry Kelts (New York: Cambridge University Press, 2006), xxi. I agree with Hutchinson (*A Treatise on Limnology*, 47) that the large number of lakes present today is "a quite exceptional phenomenon."

Although beavers appeared in the Eocene, more than 30 million years ago, apparently only the modern North American and Eurasian beavers build dams and therefore make ponds. Until people introduced beavers into Patagonia in

1946 to provide a local fur industry, all beavers, living and fossil, lived in the Northern Hemisphere. For details, see Tessa Plint, Fred J. Longstaffe, Ashley Ballantyne, Alice Telka, and Natalia Rybczynski, "Evolution of Woodcutting Behaviour in Early Pliocene Beaver Driven by Consumption of Woody Plants," *Scientific Reports* 10, no. 1 (2020): 13111, and Christopher B. Anderson, Nicolás Soto, José Luis Cabello, Guillermo Martínez Pastur, María Vanessa Lencinas, Petra K. Wallem, Daniel Antúnez et al., "*Castor canadensis* Kuhl (North American Beaver)," in *A Handbook of Global Freshwater Invasive Species*, ed. Robert A. Francis (New York: Earthscan, 2012), 277–89.

10. Information in this paragraph pertaining to water content (porosity) and water movement (permeability) comes from R. Allen Freeze and John A. Cherry, *Groundwater* (Englewood Cliffs, NJ: Prentice-Hall, 1979), 145–66, and Luc Aquilina, Christine Stumpp, Daniele Tonina, and John M. Buffington, "Hydrodynamics and Geomorphology of Groundwater Environments," in *Groundwater Ecology and Evolution*, ed. Florian Malard, Christian Griebler, and Sylvie Rétaux (London: Academic Press, 2023), 3–37.

11. *What's My Line?* was a popular game show that aired on American television from 1950 to 1975 in which panelists asked questions of guests in an attempt to guess their professions. In *What's My Lake?*, which currently airs only in my head, participants ask questions to try to guess what lives in a lake.

CHAPTER 4

1. Minnesota is commonly called the "Land of 10,000 Lakes." According to John Downing ("Minnesota: Land of How Many Lakes?," https://seagrant.umn.edu /news-info/directors-column/minnesota-land-how-many-lakes, May 17, 2021), Minnesota today contains 14,380 lakes larger than 10 acres (4 hectares), down from a high of 4.6 million lakes and ponds of all sizes right after the glaciers left.

2. Olga M. Kozhova and Lubov' R. Izmest'eva, eds., *Lake Baikal: Evolution and Biodiversity* (Leiden: Backhuys, 1998).

3. Stephanie E. Hampton, Suzanne McGowan, Ted Ozersky, Salvatore G. P. Virdis, Tuong Thuy Vu, Trisha L. Spanbauer, Benjamin M. Kraemer et al., "Recent Ecological Change in Ancient Lakes," *Limnology and Oceanography* 63, no. 5 (2018): 2277–304; "Ancient Lake," *Wikipedia*, https://en.wikipedia.org/w/index .php?title=Ancient_lake&oldid=1165204037, accessed October 5, 2023.

4. Lake Zaysan is a large lake in eastern Kazakhstan. The claims that it is 65–70 million years old appear not to be generally recognized by the scientific community, but see Spencer G. Lucas, Robert J. Emry, Viacheslav Chkhikvadze,

Bolat Bayshashov, Lyubov A. Tyutkova, Pyruza A. Tleuberdina, Ayzhan Zha-mangara et al., "Upper Cretaceous-Cenozoic Lacustrine Deposits of the Zay-san Basin, Eastern Kazakhstan," in *Lake Basins Through Space and Time*, ed. Elizabeth H. Gierlowski-Kordesch and K. R. Kelts (Tulsa: American Associ-ation of Petroleum Geologists, 2000); Benjamin F. Dorfman, "Zaysan—the Only Surviving Cretaceous Lake—May Be Lost," *Procedia Environmental Sci-ences* 10, pt. B (2011): 1376–82.

5. Although any number of websites confidently assert the ages of the oldest rivers of the world, it is tricky to estimate the age of river channels. To age riv-ers, geologists rely on things such as the age of landscape features (especially mountains) through which rivers flow, age of rock strata underlying rivers, and so on, but these various approaches often produce controversy and uncertainty about just how old a river is. For a detailed discussion of the history of the Finke River and the landscape it flows through, see Victor R. Baker, "Land-scape Evolution in the Finke (Larapinta) River Transverse Drainage, Central Mountain Ranges, Australia," *Studia Geomorphologica Carpatho-Balcanica* 55 (2021): 9–44. Rather than saying that the Finke is the oldest river in the world and that it is 350–400 million years old, it would be more accurate to say that parts of the Finke are very old and may be 300 million years old or older.

6. The New River is often said to be the second-, third-, or fourth-oldest river in the world, at ~300 million years old, but keep in mind the caveats about the difficulties of determining the age of rivers (e.g., "List of Rivers by Age," *Wiki-pedia*, https://en.wikipedia.org/w/index.php?title=List_of_rivers_by_age&ol-did=1160251868, accessed October 9, 2023). As with the Finke River, there is considerable uncertainty about its exact age (R. F. Fonner, "West Virginia Earth Science Studies: Geology of the New River Gorge," 1999, http://www.wvgs.wvnet.edu/www/geology/geoles01.htm). But it's likely that the New is old, if you know what I mean.

7. Koen Martens, "Speciation in Ancient Lakes," *Trends in Ecology and Evolution* 12, no. 5 (1997): 177–82.

8. Martens, "Speciation in Ancient Lakes," 178.

9. I here draw on the data used by Yvonne Vadeboncoeur, Peter B. McIntyre, and M. Jake Vander Zanden for their article "Borders of Biodiversity: Life at the Edge of the World's Large Lakes" (*BioScience* 61, no. 7 [2011]: 526–37), which Vadeboncoeur kindly provided to me.

10. Vincent Kotwicki and Robert Allan, "La Niña de Australia: Contemporary and Palaeo-Hydrology of Lake Eyre," *Palaeogeography, Palaeoclimatology, Palaeoecology*

144, nos. 3–4 (1998): 265–80; Anna Habeck-Fardy and Gerald C. Nanson, "Environmental Character and History of the Lake Eyre Basin, One-Seventh of the Australian Continent," *Earth-Science Reviews* 132 (2014): 39–66.

11.   Andrea Belgrano and Charles W. Fowler, "How Fisheries Affect Evolution," *Science* 342, no. 6163 (2013): 1176–77, and references cited therein.

12.   Science for Environment Policy, *Synthetic Biology and Biodiversity* (Bristol, UK: Science Communication Unit, University of the West of England, 2016), https://data.europa.eu/doi/10.2779/976543; Eleonore Pauwels, "The Rise of Citizen Bioscience," *Scientific American Observations*, January 5, 2018, https://blogs.scientificamerican.com/observations/the-rise-of-citizen-bioscience/.

CHAPTER 5

1.   William D. Williams, Patrick De Deckker, and Russel J. Shiel, "The Limnology of Lake Torrens, an Episodic Salt Lake of Central Australia, with Particular Reference to Unique Events in 1989," *Hydrobiologia* 384, no. 1 (1998): 101–10; Vincent Kotwicki and Robert Allan, "La Niña de Australia: Contemporary and Palaeo-Hydrology of Lake Eyre," *Palaeogeography, Palaeoclimatology, Palaeoecology* 144, nos. 3–4 (1998): 265–80; Anna Habeck-Fardy and Gerald C. Nanson, "Environmental Character and History of the Lake Eyre Basin, One-Seventh of the Australian Continent," *Earth-Science Reviews* 132 (2014): 39–66.

2.   This account of Bretz's life and the Lake Missoula floods is based on J Harlen Bretz, "The Channeled Scablands of the Columbia Plateau," *Journal of Geology* 31, no. 8 (1923): 617–89, Jim E. O'Connor and John E. Costa, "The World's Largest Floods, Past and Present: Their Causes and Magnitudes," *United States Geological Survey Circular* 1254 (2004): 1–13, Cassandra Tate, "Bretz, J Harlen, 1882–1981," *HistoryLink.org*, November 29, 2007, https://www.historylink.org/File/8382, John Soennichsen, "Legacy: J Harlen Bretz (1882?1981)," *University of Chicago Magazine* (November–December 2009), http://magazine.uchicago.edu/0912/features/legacy.shtml, Jim E. O'Connor, Victor R. Baker, Richard B. Waitt, Larry N. Smith, Charles M. Cannon, David L. George, and Roger P. Denlinger, "The Missoula and Bonneville Floods—A Review of Ice-Age Megafloods in the Columbia River Basin," *Earth-Science Reviews* 208 (2020): 103181, and "Missoula Floods," *Wikipedia*, https://en.wikipedia.org/w/index.php?title=Missoula_floods&oldid=1165864966, accessed October 10, 2023.

3.   Victor R. Baker, Gerardo Benito, and Alexey N. Rudoy, "Paleohydrology of Late Pleistocene Superflooding, Altay Mountains, Siberia," *Science* 259, no. 5093 (1993): 348–50; O'Connor and Costa, "The World's Largest Floods, Past and Present."

CHAPTER 6

1. Jean-Pierre Gattuso, Alexandre Magnan, Raphaël Billé, William W. L. Cheung, Ella L. Howes, Fortunat Joos, Denis Allemand et al., "Contrasting Futures for Ocean and Society From Different Anthropogenic CO_2 Emissions Scenarios," *Science* 349, no. 6243 (2015): aac4722; Scott C. Doney, D. Shallin Busch, Sarah R. Cooley, and Kristy J. Kroeker, "The Impacts of Ocean Acidification on Marine Ecosystems and Reliant Human Communities," *Annual Review of Environment and Resources* 45 (2020): 83–112.

2. It's hard to say exactly what the pH of the most acidic natural lake is because it is tricky to measure or estimate pHs below about 0.5. Nevertheless, the pH of Kawah Ijen has repeatedly been reported to be 0.3, and pHs as low as slightly negative have been reported from volcanic lakes; see Johan C. Varekamp, Gregory B. Pasternack, and Gary L. Rowe Jr., "Volcanic Lake Systematics," pt. 2, "Chemical Constraints," *Journal of Volcanology and Geothermal Research* 97, no. 1 (2000): 161–79, and Johan C. Varekamp, "The Chemical Composition and Evolution of Volcanic Lakes," in *Volcanic Lakes*, edited by Dmitri Rouwet et al. (Berlin: Springer, 2015), 93–123. It seems safe to guess that a pH in the neighborhood of 0.1 is about right for the most acidic of these waters.

3. Ansje J. Lohr, Rutger Sluik, Mary M. Olaveson, Núria Ivorra, C. V. van Gestel, and N. V. van Straalen, "Macroinvertebrate and Algal Communities in an Extremely Acidic River and the Kawah Ijen Crater Lake (pH< 0.3), Indonesia," *Archiv für Hydrobiologie* 165, no. 1 (2006): 1–21.

4. Christa Schleper, Gabriela Puehler, Ingelore Holz, Agata Gambacorta, Davorin Janekovic, Ute Santarius, Hans-Peter Klenk et al., "*Picrophilus* Gen. Nov., Fam. Nov.: a Novel Aerobic, Heterotrophic, Thermoacidophilic Genus and Family Comprising Archaea Capable of Growth Around pH 0," *Journal of Bacteriology* 177, no. 24 (1995): 7050–59.

5. D. Kirk Nordstrom and Charles N. Alpers, "Negative pH, Efflorescent Mineralogy, and Consequences for Environmental Restoration at the Iron Mountain Superfund Site, California," *Proceedings of the National Academy of Sciences* 96 (1999): 3455–62.

6. Ekkehard Vareschi, "The Ecology of Lake Nakuru (Kenya)," pt. 3, "Abiotic Factors and Primary Production," *Oecologia* 55, no. 1 (1982): 81–101; Ekkehard Vareschi and Jim L. Jacobs, "The Ecology of Lake Nakuru," pt. 4, "Synopsis of Production and Energy Flow," *Oecologia* 65, no. 3 (1985): 412–24.

7. The Yellow River often contains more than a pound of sediment per gallon of "water" (>100 kilograms per cubic meter). The highest value I found was for one of its tributaries, where the river contained more than 13 pounds of

sediment per gallon (1600 kilograms per cubic meter), resulting in a fluid that was 60% sediment by volume (Ning Chien and Zhaohui Wan, *Mechanics of Sediment Transport* [Reston, VA: ASCE Press, 1999]).

8. Contrary to the claims of tourism offices and civic boosters around the world, the clearest inland water yet found appears to be Blue Lake, a remote, mountain lake on the South Island of New Zealand, where scientists recorded an average water clarity of 74.2 meters (243 feet), with one reading as high as 81.4 meters (267 feet). The theoretical maximum clarity for pure water is about 80 meters (262 feet), so it is unlikely that waters much clearer than Blue Lake will be discovered. For details, see Mark P. Gall, Rob J. Davies-Colley, and Rob A. Merrilees, "Exceptional Visual Clarity and Optical Purity in a Sub-Alpine Lake," *Limnology and Oceanography* 58, no. 2 (2013): 443–51.

CHAPTER 7

1. These are characters from the American television comedy *Gilligan's Island*, which ran from 1964 to 1967 and has been widely rerun since then. The show follows an eclectic group of 7 people who are shipwrecked onto a tropical island in the Pacific, as they try to survive and escape. According to Wikipedia, the characters of this show are now (God help us) "cultural icons" ("Gilligan's Island," *Wikipedia*, https://en.wikipedia.org/w/index.php?title=Gilligan%27s _Island&oldid=1179583771, accessed October 12, 2023).

2. Sherwin Carlquist, *Island Biology* (New York: Columbia University Press, 1974); David W. Steadman, *Extinction and Biogeography of Tropical Pacific Birds* (Chicago: University of Chicago Press, 2006); Jonathan B. Losos and Robert E. Ricklefs, "Adaptation and Diversification on Islands," *Nature* 457 (2009): 830–36; Patrick O'Grady and Rob DeSalle, "Hawaiian *Drosophila* as an Evolutionary Model Clade: Days of Future Past," *BioEssays* 40, no. 5 (2018): 1700246.

3. Bernie R. Tershy, Kuo-Wei Shen, Kelly M. Newton, Nick D. Holmes, and Donald A. Croll, "The Importance of Islands For the Protection of Biological and Linguistic Diversity," *BioScience* 65 (2015): 592–97, James C. Russell and Christoph Kueffer, "Island Biodiversity in the Anthropocene," *Annual Review of Environment and Resources* 44 (2019): 31–60.

4. John A. Downing, Yves T. Prairie, Jonathan J. Cole, Carlos M. Duarte, Lars J. Tranvik, Robert G. Striegl, William H. McDowell et al., "The Global Abundance and Size Distribution of Lakes, Ponds, and Impoundments," *Limnology and Oceanography* 51, no. 5 (2006): 2388–97, "Global Island Database," http://www.globalislands.net/about/gid_functions.php, accessed October 10, 2023.

CHAPTER 8

1. Christian Lévêque, Thierry Oberdorff, Didier Paugy, Melanie L. J. Stiassny, and P. A. Tedesco, "Global Diversity of Fish (Pisces) in Freshwater," in *Freshwater Animal Diversity Assessment*, ed. Estelle V. Balian, Christian Lévêque, Hendrik H. Segers, and Koen Martens (Dordrecht: Springer, 1998), 545–67; Estelle V. Balian, Hendrik H. Segers, Koen Martens, and Christian Lévêque, "The Freshwater Animal Diversity Assessment: An Overview of the Results," in *Freshwater Animal Diversity Assessment*, 627–37; Greta Carrete Vega and John J. Wiens, "Why Are There So Few Fish in the Sea?," *Proceedings of the Royal Society B* 279 (2012): 2323–29.

2. Patricia A. Chambers, Paresh Lacoul, Kevin J. Murphy, and Sidinei Magela Thomaz, "Global Diversity of Aquatic Macrophytes in Freshwater," in *Freshwater Animal Diversity Assessment*, 9–26.

3. Walter K. Dodds and Matt R. Whiles, *Freshwater Ecology: Concepts and Environmental Applications of Limnology*, 2nd ed. (Burlington, VT: Academic Press, 2010), 200; Olivier De Clerck, Michael D. Guiry, Frederik Leliaert, Yves Samyn, and Heroen Verbruggen, "Algal Taxonomy: A Road to Nowhere?," *Journal of Phycology* 49, no. 2 (2013): 215–25.

4. Based on Christopher Scarpf, compiler, "Checklist of Freshwater Fishes of North America, Including Subspecies and Undescribed Forms," http://www.nanfa.org/checklist.shtml, accessed November 30, 2022.

5. See *Freshwater Animal Diversity Assessment*.

6. Renata Manconi and Roberto Pronzato, "Global Diversity of Sponges (Porifera: Spongillina) in Freshwater," in *Freshwater Animal Diversity Assessment*, 27–33.

7. Thomas Jankowski, "The Freshwater Medusae of the World—A Taxonomic and Systematic Literature Study With Some Remarks On Other Inland Water Jellyfish," *Hydrobiologia* 462, nos. 1–3 (2001): 91–113; Thomas Jankowski, Allen G. Collins, and Richard Campbell, "Global Diversity of Inland Water Cnidarians," in *Freshwater Animal Diversity Assessment*, 35–40.

8. Geraldine Veron, Bruce D. Patterson, and Randall Reeves, "Global Diversity of Mammals (Mammalia) in Freshwater," in *Freshwater Animal Diversity Assessment*, 607–17; David Dudgeon, *Threatened Freshwater Animals of Tropical East Asia* (New York: Routledge, 2023), 302–8.

9. Samuel T. Turvey, Robert L. Pitman, Barbara L. Taylor, Jay Barlow, Tomonari Akamatsu, Leigh A. Barrett, Xiujiang Zhao et al., "First Human-Caused Extinction of a Cetacean Species?," *Biology Letters* 3, no. 5 (2007): 537–40; Dudgeon, *Threatened Freshwater Animals of Tropical East Asia*, 306.

10. Some species of marine sharks, most notably the bull shark, stray far into fresh water—a bull shark was captured in the Mississippi River at Alton, Illinois, 1,150 miles (1,850 kilometers) upriver from the Gulf of Mexico (Ryan Shell, Nicholas Gardner, and Robert A. Hrabik, "Updates on Putative Bull Shark (*Carcharhinus leucas*) Occurrences in the Upper Mississippi River Basin of North America," *Marine and Fishery Sciences* 36, no. 1 [2023]: 91–100). However, it is doubtful whether there are any truly freshwater sharks that can complete their life cycles in fully fresh water. Dudgeon (*Threatened Freshwater Animals of Tropical East Asia*, 174–75) discusses the evidence that the critically endangered Ganges river shark may be a true freshwater species. But I guess if you're eaten by a shark when you're swimming in a freshwater river or lake, it probably doesn't much matter to you whether it is a truly freshwater shark or just a wandering marine shark.

11. Joe DiMaggio was a famous American baseball player, and the arcane terms I use here refer to baseball plays.

12. Maurice Kottelat, Ralf Britz, Tan Heok Hui, and Kai-Erik Witte, "*Paedocypris*, a New Genus of Southeast Asian Cyprinid Fish with a Remarkable Sexual Dimorphism, Comprises the World's Smallest Known Vertebrate," *Proceedings of the Royal Society B* 273 (2006): 895-99.

13. Roger Bour, "Global Diversity of Turtles (Chelonii; Reptilia) in Freshwater," in *Freshwater Animal Diversity Assessment*, 593–98.

14. Samuel Martin, "Global Diversity of Crocodiles (Crocodilia, Reptilia) in Freshwater," in *Freshwater Animal Diversity Assessment*, 587–91; Gordon Grigg and David Kirshner, *Biology and Evolution of Crocodylians* (Ithaca, NY: Cornell University Press, 2015); Dudgeon, *Threatened Freshwater Animals of Tropical East Asia*, 233–41.

15. The many species of "terrestrial" birds and mammals that live most of their lives above the water's surface but that depend to a greater or lesser degree on inland waters for habitat, food, and other necessities often are excluded from tallies of aquatic diversity. However, populations of many of these species decline or disappear when the aquatic habitats are destroyed or damaged, so they must be counted in some way as parts of inland-water ecosystems. Dudgeon (*Threatened Freshwater Animals of Tropical East Asia*, 270–320) does a nice job listing these many species and describing how they depend on (and influence) the rivers of tropical Asia. See also Olivier Dehorter and Matthieu Guillemain, "Global Diversity of Freshwater Birds," in *Freshwater Animal Diversity Assessment*, 619–26, and Veron et al., "Global Diversity of Mammals (Mammalia) in Freshwater."

16. Lévêque et al., "Global Diversity of Fish (Pisces) in Freshwater."

17. Julien Cucherousset, Stéphanie Boulêtreau, Frédéric Azémar, Arthur Compin, Mathieu, Guillaume, and Frédéric Santoul, " 'Freshwater Killer Whales': Breaching Behavior of an Alien Fish to Hunt Land Birds," *PLoS One* 7, no. 12 (2012): e50840. This article includes a memorable movie of these big catfish at work.

18. Maurice Kottelat gives a maximum weight of 350 kilograms (770 pounds) (*Fishes of Laos* [Colombo: Wildlife Heritage Trust of Sri Lanka, 2001], 130).

19. Exactly 18 specimens of this tiny catfish were collected between 1943 and 1957, 17 of them from a 20-foot (6-meter) section of a single riffle in Big Darby Creek. Despite concerted efforts to find this species at this site or anywhere else, no further specimens have been found since 1957, and the species is now thought to be extinct (Milton B. Trautman, *The Fishes of Ohio*, revised ed. [Columbus: Ohio State University Press, 1981], 503–5; US Fish and Wildlife Service, "Endangered and Threatened Wildlife and Plants; Removal of 23 Extinct Species from the Lists of Endangered and Threatened Wildlife and Plants," *Federal Register* 86 [2021]: 54318–19).

20. When asked what we might be able to infer about God from the study of natural history, J. B. S. Haldane replied: "God has an inordinate fondness for beetles."

21. As far as I can tell, the largest inland-water invertebrate is the Tasmanian crayfish (fig. 8.6, *bottom*) at about 7 pounds (3 kilograms) wet weight, and the smallest must be one of the smaller rotifers, roundworms, or gastrotrichs, which probably tip the scales at about one ten-billionth of a pound (0.04 micrograms), wet weight (David Strayer, "The Size Structure of a Lacustrine Zoobenthic Community," *Oecologia* 69, no. 4 [1986]: 513–16).

22. "Kaipen," *Wikipedia,* https://en.wikipedia.org/w/index.php?title=Kaipen&oldid=1142150416, accessed October 13, 2023.

23. G. Evelyn Hutchinson, *A Treatise on Limnology,* 4 vols. (New York: Wiley, 1957–93), 3:2.

24. Wayne W. Carmichael, Sandra M. F. O. Azevedo, Ji Si An, Renato J. R. Molica, Elise M. Jochimsen, Sharon Lau, and Kenneth L. Rinehart et al., "Human Fatalities from Cyanobacteria: Chemical and Biological Evidence for Cyanotoxins," *Environmental Health Perspectives* 109, no. 7 (2001): 663–68; Michael Wines, "Behind Toledo's Water Crisis, a Long-Troubled Lake Erie," *New York Times,* August 4, 2014, A12.

25. The rule of thumb among aquatic microbial ecologists is that bacterial densities in surface waters are typically on the order of millions per cubic centimeter of water and billions per cubic centimeter of sediment.

26.   Biologists are still trying to estimate the number of parasite species on Earth,
       but it's entirely possible that half of the animal species on the planet are
       parasites. For examples of bold attempts to evaluate global parasite species
       richness from woefully inadequate data, see Andy Dobson, Kevin D. Lafferty,
       Armand M. Kuris, Ryan F. Hechinger, and Walter Jetz, "Homage to Linnaeus:
       How Many Parasites? How Many Hosts?," *Proceedings of the National Academy
       of Sciences* 105, supp. (2008): 11482–89, Robert Poulin and Serge Morand, *Par-
       asite Biodiversity* (Washington, DC: Smithsonian Institution, 2004), Colin J.
       Carlson, Tad A. Dallas, Laura W. Alexander, Alexandra L. Phelan, and Anna J.
       Phillips, "What Would It Take To Describe the Global Diversity of Parasites?,"
       Proceedings of the Royal Society B 287 (2020): 20201841.

27.   Luke Tain, Marie-Jeanne Perrot-Minnot, and Frank Cézilly, "Altered Host
       Behaviour and Brain Serotonergic Activity Caused by Acanthocephalans: Ev-
       idence For Specificity," *Proceedings of the Royal Society B* 273 (2006): 3039–45.
       For a general review of manipulation of host behaviors by parasites, see Rob-
       ert Poulin, "Parasite Manipulation of Host Behavior: An Update and Fre-
       quently Asked Questions," in *Advances in the Study of Behavior,* vol. 41, ed. H.
       Jane Brockmann (Burlington, VT: Academic Press, 2010), 151–86.

28.   The basic life cycles of these alarming parasites (horsehair worms [Nema-
       tomorpha] and mermithid nematodes) are described in standard textbooks
       of invertebrate zoology, such as George O. Poinar Jr., "Phylum Nemata," in
       Thorp and Covich's Freshwater Invertebrates, vol. 1: *Ecology and General Biology,*
       4th ed., ed. James H. Thorp and D. Christopher Rogers (London: Academic
       Press, 2015), 273–302, and Matthew G. Bolek, Andreas Schmidt-Rhaesa, L.
       Cristina De Villalobos, and Ben Hanelt, "Phylum Nematomorpha," in *Thorp
       and Covich's Freshwater Invertebrates,* 1:303–26. See also Frédéric Thomas,
       Andreas Schmidt-Rhaesa, Guilhaume Martin, C. Manu, Patrick Durand, and
       Florent Renaud, "Do Hairworms (Nematomorpha) Manipulate the Water
       Seeking Behaviour of Their Terrestrial Hosts?," *Journal of Evolutionary Biol-
       ogy* 15, no. 3 (2002): 356–61; Jean-François Doherty and Robert Poulin, "The
       Return to Land: Association Between Hairworm Infection and Aquatic Insect
       Development," *Parasitology Research* 121, no. 2 (2022): 667–73.

29.   If you're not squeamish, you can find basic information about these human
       parasites in standard textbooks (e.g., Burton J. Bogitsh, Clint E. Carter, and
       Thomas N. Oeltmann, *Human Parasitology,* 5th ed. [London: Academic Press,
       2019]) or in "Human Parasite," *Wikipedia,* https://en.wikipedia.org/w/index
       .php?title=Human_parasite&oldid=1143975595, and the links included therein,
       accessed October 17, 2023.

30.   Numbers 21:6 ("And the Lord sent fiery serpents among the people, and they bit the people"). Some scholars interpret the "fiery serpent" as the guinea worm, but this interpretation is not universally accepted; see David I. Grove, *A History of Human Helminthology* (Wallingford, UK: CAB International, 1990), 694.

31.   Peter J. Hudson, Andrew P. Dobson, and Kevin D. Lafferty, "Is a Healthy Ecosystem One That Is Rich in Parasites?," *Trends in Ecology and Evolution* 21, no. 7 (2006): 381–85; Eric R. Dougherty, Colin J. Carlson, Veronica M. Bueno, Kevin R. Burgio, Carrie A. Cizauskas, Christopher F. Clements, Dana P. Seidel et al., "Paradigms For Parasite Conservation," *Conservation Biology* 30, no. 4 (2015): 724–33; Colin J. Carlson, Kevin R. Burgio, Eric R. Dougherty, Anna J. Phillips, Veronica M. Bueno, Christopher F. Clements, Giovanni Castaldo et al., "Parasite Biodiversity Faces Extinction and Redistribution in a Changing Climate," *Science Advances* 3, no. 9 (2017): e1602422; Colin J. Carlson, Skylar Hopkins, Kayce C. Bell, Jorge Doña, Stephanie S. Godfrey, Mackenzie L. Kwak, Kevin D. Lafferty et al., "A Global Parasite Conservation Plan," *Biological Conservation* 250 (2020): 108596.

32.   Dougherty et al., "Paradigms For Parasite Conservation"; Carlson et al., "Parasite Biodiversity Faces Extinction and Redistribution in a Changing Climate"; Carlson et al., "A Global Parasite Conservation Plan."

CHAPTER 9

1.   The longest distance that I found for salmon was 3,204 kilometers, for Chinook salmon in the Yukon River headwaters (Randy J. Brown, Al von Finster, Robert J. Henszey, and John H. Eiler, "Catalog of Chinook Salmon Spawning Areas in Yukon River Basin in Canada and United States," *Journal of Fish and Wildlife Management* 8, no. 2 [2017]: 558–86). Impressive, yes, but one of the goliath catfishes of the Amazon migrates 3,400 miles (5,800 kilometers) between the mouth of the river and the fish's spawning grounds in the Andes (Ronaldo B. Barthem, Michael Goulding, Rosseval G. Leite, Carlos Cañas, Bruce Forsberg, Eduardo Venticinque, and Paulo Petry, "Goliath Catfish Spawning in the Far Western Amazon Confirmed by the Distribution of Mature Adults, Drifting Larvae, and Migrating Juveniles," *Scientific Reports* 7 [2017]: 41784).

2.   Migratory fishes have many functions in stream ecosystems, and when these fishes are lost, those functions are no longer carried out. For instance, salmon in the Pacific Northwest carry large amounts of nutrients in their bodies upriver to their spawning beds, where these nutrients feed birds, bears, and stream insects and even fertilize streamside forests. Other migratory fishes

scour streambeds, which affects sediment deposition and algal production in the stream, and transport of sediments and nutrients downriver. For a good summary, see Alexander S. Flecker, Peter B. McIntrye, Jonathan W. Moore, Jill T. Anderson, Brad W. Taylor, and Robert O. Hall Jr., "Migratory Fishes as Material and Process Subsidies in Riverine Ecosystems," in *Community Ecology of Stream Fishes: Concepts, Approaches, and Techniques*, ed. Keith B. Gido and Donald A. Jackson (Bethesda, MD: American Fisheries Society, 2010), 559–92.

3.   NOAA has a great website that shows radar maps of massive mayfly emergences along the upper Mississippi River, along with impressive photographs taken from ground level of these swarms of emerging insects ("Tracking Mayflies," https://www.weather.gov/arx/mayfly_tracking, accessed October 18, 2023).

4.   Michael L. May, "Dispersal by Aquatic Insects," in *Aquatic Insects*, ed. Kleber Del-Claro and Rhainer Guillermo (Cham: Springer, 2019), 35–73.

5.   H. B. N. Hynes, *The Ecology of Running Waters* (Toronto: University of Toronto Press, 1970), 153.

6.   Daniel W. Schneider and Thomas M. Frost, "Massive Upstream Migrations by a Tropical Freshwater Neritid Snail," *Hydrobiologia* 137 (1986): 153–57; Yasunori Kano and Hiroaki Fukumori, "Neritidae Rafinesque, 1815," in *Freshwater Mollusks of the World: A Distribution Atlas*, ed. Charles Lydeard and Kevin S. Cummings (Baltimore, MD: Johns Hopkins University Press, 2019), 31–36. Don't you love the thought of "massive" snail migrations? Like tiny wildebeests moving across the Serengeti, but in (very) slow motion.

7.   M. Christopher Barnhart, Wendell R. Haag, and William N. Roston, "Adaptations to Host Infection and Larval Parasitism in Unionoida," *Journal of the North American Benthological Society* 27 (2008): 370–394.

8.   Wendell R. Haag, Robert S. Butler, and Paul D. Hartfield, "An Extraordinary Reproductive Strategy in Freshwater Bivalves: Prey Mimicry to Facilitate Larval Dispersal," *Freshwater Biology* 34, no. 2 (1995): 471–76.

9.   Barnhart et al., "Adaptations to Host Infection and Larval Parasitism in Unionoida," 379.

10.   Ragnar Kinzelbach, "The Main Features of the Phylogeny and Dispersal of the Zebra Mussel *Dreissena polymorpha*," in *The Zebra Mussel Dreissena polymorpha*, vol. 4, ed. Dietrich Neumann and Henk A. Jenner (Stuttgart: Gustav Fischer, 1992), 5–17; Bart J. A. Pollux, Gerard van der Velde, and Abraham bij de Vaate, "A Perspective on the Global Spread of *Dreissena polymorpha*: A Review on Possibilities and Limitations," in *The Zebra Mussel in Europe*, ed. Gerard van

de Velde, Sanjeevi Rajagopal, and Abraham bij de Vaate (Leiden: Backhuys, 2010), 45–58.

11. C. Thomas Philbrick, "Aspects of Floral Biology, Breeding System, and Seed and Seedling Biology in *Podostemum ceratophyllum* (Podostemaceae)," *Systematic Biology* 9, no. 2 (1984): 166–74; C. Thomas Philbrick, Paula K. B. Philbrick, and Brandon M. Lester, "Root Fragments as Dispersal Propagules in the Aquatic Angiosperm *Podostemum ceratophyllum* Michx. (Hornleaf Riverweed, Podostemaceae)," *Northeastern Naturalist* 22, no. 3 (2015): 643–47.

12. C. Thomas Philbrick and Alejandro Novelo R., "New World Podostemaceae: Ecological and Evolutionary Enigmas," *Brittonia* 47, no. 2 (1995): 210–22; C. Thomas Philbrick and Alejandro Novelo R., "Monograph of *Podostemum* (Podostemaceae)," *Systematic Botany Monographs* 70 (2004): 1–106.

13. Christo Vladimirov Javacheff and Jeanne-Claude Denat de Guillebon, working under the name Christo, were artists whose best-known works involved wrapping large parts of the landscape in bright fabrics. "Surrounded Islands" is a good example; see https://christojeanneclaude.net/artworks /surrounded-islands/.

14. David D. Hart and Christopher M. Finelli, "Physical-Biological Coupling in Streams: The Pervasive Effects of Flow on Benthic Organisms," *Annual Review of Ecology and Systematics* 30 (1999): 363–95.

15. Victor Kang, Robin T. White, Simon Chen, and Walter Federle, "Extreme Suction Attachment Performance from Specialised Insects Living in Mountain Streams (Diptera: Blepharoceridae)," *eLife* 10 (2021): e63250.

16. Jürgen Weissenberger, Hugo Spatz, Anneliese Emmans, and Jürgen Schwoerbel, "Measurement of Lift and Drag Forces in the mN Range Experienced by Benthic Arthropods at Flow Velocities Below 1.2 m s^{-1}," *Freshwater Biology* 25 (1991): 21–31; Hart and Finelli, "Physical-Biological Coupling in Streams."

17. This unusual experiment was reported by Milton B. Trautman, *The Fishes of Ohio*, revised ed. (Columbus: Ohio State University Press, 1981), 457.

18. G. Evelyn Hutchinson, *A Treatise on Limnology*, 4 vols. (New York: Wiley, 1957–93), 2:188–90; Robert W. Pennak, "The Fresh-Water Invertebrate Fauna: Problems and Solutions for Evolutionary Success," *American Zoologist* 25, no. 3 (1985): 671–87.

19. John C. McNamara and Carolina Arruda Freire, "Strategies of Invertebrate Osmoregulation: An Evolutionary Blueprint for Transmuting into Fresh Water from the Sea," *Integrative and Comparative Biology* 62, no. 2 (2022): 376–87; Hutchinson, *A Treatise on Limnology*, 2:188–90; Pennak, "The Fresh-Water Invertebrate Fauna."

CHAPTER 10

1.   But it's not impossible for even small creatures to walk or crawl to safety. David A. Lytle, Julian D. Olden, and Laura E. McMullen, "Drought-Escape Behaviors of Aquatic Insects May Be Adaptations to Highly Variable Flow Regimes Characteristic of Desert Rivers," *Southwestern Naturalist* 53, no. 3 (2008): 399–402, provide a striking account (with pictures) of a 120-foot-long (37-meter) column of beetles and dragonfly larvae marching upstream to find water as their small stream dried up.

2.   Kenneth J. Kingsley, "*Eretes sticticus* (L.) (Coleoptera: Dytiscidae): Life History Observations and an Account of a Remarkable Event of Synchronous Emigration from a Temporary Desert Pond," *Coleopterists Bulletin* 39 (1985): 7–10.

3.   Michael L. May, "Dispersal in Aquatic Insects," in *Aquatic Insects: Behavior and Ecology*, ed. Kleber Del-Claro and Rhainer Guillermo (Cham: Springer, 2019), 43.

4.   Kenneth J. Boss, "Oblomovism in the Mollusca," *Transactions of the American Microscopical Society* 93, no. 4 (1974): 460–81.

5.   Alfred P. Fishman, Allan I. Pack, Richard G. Delaney, and Raymond J. Galante, "Estivation in *Protopterus*," *Journal of Morphology* 190, supp. 1 (1986): 237–48; Ai M. Loong, Cheryl Y. M. Pang, Kum C. Hiong, Wai P. Wong, Shit F. Chew, and Yuen K. Ip, "Increased Urea Synthesis and/or Suppressed Ammonia Production in the African Lungfish, *Protopterus annectens*, during Aestivation in Air or Mud," *Journal of Comparative Physiology B* 178, no. 3 (2008): 351–63; Ryan D. Heimroth, Elisa Casadei, Ottavia Benedicenti, Chris Tsuyoshi Amemiya, Pilar Muñoz, and Irena Salinas, "The Lungfish Cocoon is a Living Tissue with Antimicrobial Functions," *Science Advances* 7, no. 47 (2021): eabj0829.

6.   Information about bdelloids, tardigrades, and their remarkable abilities to shut down in the face of adversity can be found in John H. Crowe, Folkert A. Hoekstra, and Lois M. Crowe, "Anhydrobiosis," *Annual Review of Physiology* 54 (1992): 579–99, Lorena Rebecchi, Tiziana Altiero, and Roberto Guidetti, "Anhydrobiosis: The Extreme Limit of Desiccation Tolerance," *Invertebrate Survival Journal* 4, no. 2 (2007): 65–81, Diane R. Nelson, Roberto Guidetti, and Lorena Rebecchi, "Phylum Tardigrada," in *Thorp and Covich's Freshwater Invertebrates*, vol. 1, *Ecology and General Biology*, 4th ed., ed. James H. Thorp and D. Christopher Rogers (London: Academic Press, 2015), 361–63, Robert L. Wallace, Terry W. Snell, and Hilary A. Smith, "Phylum Rotifera," in *Thorp and Covich's Freshwater Invertebrates*, 1:240–42, and David A. Wharton, "Anhydrobiosis," *Current Biology* 25, no. 23 (2015): R1114–R1116.

7.   "It may be doubted, however, to what extent the word living may be properly applied to organisms that seem rather to be reversibly dead" (G. Evelyn Hutchinson, *A Treatise on Limnology*, 4 vols. [New York: Wiley, 1957–93], 2:512).

8.   The list of the environmental tolerances of tardigrades comes from Nelson et al., "Phylum Tardigrada," and Alejandra Traspas and Mark J. Burchell, "Tardigrade Survival Limits in High-Speed Impacts—Implications for Panspermia and Collection of Samples From Plumes Emitted by Ice Worlds," *Astrobiology* 21, no. 7 (2021): 845–52.

9.   The hapless Wile E. Coyote is a character in the American *Looney Toons* and *Merrie Melodies* cartoons who repeatedly tries to catch and eat the Roadrunner, often using explosives and various improbable devices (including anvils) bought from the Acme Corporation. He never succeeds.

10. K. Ingemar Jönsson, Elke Rabbow, Ralph O. Schill, Mats Harms-Ringdahl, and Petra Rettberg, "Tardigrades Survive Exposure to Space in Low Earth Orbit," *Current Biology* 18, no. 17 (2008): R729–R731; Weronika Erdmann and Łukasz Kaczmarek, "Tardigrades in Space Research—Past and Future," *Origins of Life and Evolution of Biospheres* 47, no. 4 (2017): 545–53.

11. Robert W. Pennak, "The Fresh-Water Invertebrate Fauna: Problems and Solutions for Evolutionary Success," *American Zoologist* 25, no. 3 (1985): 671–87; Carla E. Cáceres, "Dormancy in Invertebrates," *Invertebrate Biology* 116, no. 4 (1997): 371–83.

12. J. Shen-Miller, Mary Beth Mudgett, J. William Schopf, Steven Clarke, and Rainer Berger, "Exceptional Seed Longevity and Robust Growth: Ancient Sacred Lotus from China," *American Journal of Botany* 82, no. 11 (1995): 1367–80.

13. D. Dudley Williams, *The Biology of Temporary Waters* (Oxford: Oxford University Press, 2006); Alexander D. Huryn, J. Bruce Wallace, and Norman H. Anderson, "Habitat, Life History, Secondary Production, and Behavioral Adaptations of Aquatic Insects," in *An Introduction to the Aquatic Insects of North America*, 4th ed., edited by Richard W. Merritt, Kenneth W. Cummins, and Martin B. Berg (Dubuque, IA: Kendell/Hunt Publishing Company, 2008), 55–103; May, "Dispersal in Aquatic Insects," 37.

14. Bassett Maguire Jr., "The Passive Dispersal of Small Aquatic Organisms and Their Colonization of Isolated Bodies of Water," *Ecological Monographs* 33 (1963): 161–85; Naveen K. Sharma, Ashwani Kumar Rai, Surendra Singh, and Richard Malcolm Brown Jr., "Airborne Algae: Their Present Status and Relevance," *Journal of Phycology* 43, no. 2 (2007): 615–27; Christoph Ptatscheck, Birgit Gansfort, and Walter Traunspurger, "The Extent of Wind-Mediated

Dispersal of Small Metazoans, Focusing [sic] Nematodes," *Scientific Reports* 8, no. 1 (2018): 6814.

15.   Tim M. Berra and Gerald R. Allen, "Burrowing, Emergence, Behavior, and Functional Morphology of the Australian Salamanderfish, *Lepidogalaxias salamandroides*," *Fisheries* 14, no. 5 (1989), 2–10; Joseph S. Nelson, *Fishes of the World*, 3rd ed. (New York: Wiley, 1994).

CHAPTER 11

1.   Skittles and Snickers bars are popular kinds of candy.

2.   Lakshminath Kundanati, Prashant Das, and Nicola M. Pugno, "Prey Capturing Dynamics and Nanomechanically Graded Cutting Apparatus of Dragonfly Nymph," *Materials* 14, no. 3 (2021): 559.

3.   For an introduction to the anatomy and probable habits of griffenflies, see André Nel, Jakub Prokop, Martina Pecharová, Michael S. Engel, and Romain Garrouste, "Palaeozoic Giant Dragonflies Were Hawker Predators," *Scientific Reports* 8, no. 1 (2018): 12141.

4.   Barbara L. Peckarsky, Cathy A. Cowan, Marjory A. Penton, and Chester Anderson, "Sublethal Consequences of Stream-Dwelling Predatory Stoneflies on Mayfly Growth and Fecundity," *Ecology* 74, no. 6 (1993): 1836–46.

5.   Brian D. Smith, Mya Than Tun, Aung Myo Chit, Han Win, and Thida Moe, "Catch Composition and Conservation Management of a Human-Dolphin Cooperative Cast-Net Fishery in the Ayeyarwady River, Myanmar," *Biological Conservation* 142, no. 5 (2009): 1042–49; Doug Clark, "In a Fragile Partnership, Dolphins Help Catch Fish in Myanmar," *New York Times*, September 1, 2017, A4; David Dudgeon, *Threatened Freshwater Animals of Tropical East Asia* (London: Routledge, 2023), 303–4.

6.   You can see more photographs and drawings of the varied, interesting, and sometimes scary mouthparts of rotifers at "Rotifer Trophi Web Page," http://www.rotifera.hausdernatur.at/Rotifer_data/trophi/start.html, as well as in the technical scientific literature (e.g., Willem H. de Smet and Roger Pourriot, *Rotifera*, vol. 5: *The Dicranophoridae (Monogononta): The Ituridae (Monogononta)* (Amsterdam: SPB Publishing, 1997).

7.   Information about flatworm predation is summarized from William A. Kepner and John F. Barker, "Nematocysts of *Microstoma*," *Biological Bulletin* 47, no. 4 (1924): 239–52, Carolina Noreña, Cristina Damborenea, and Francisco Brusa, "Phylum Platyhelminthes," in *Thorp and Covich's Freshwater Invertebrates*, vol. 1: *Ecology and General Biology*, 4th ed., ed. James H. Thorp and D. Christopher

Rogers (London: Academic Press, 2015), 181–203, Georg Krohne, "Organelle Survival in a Foreign Organism: *Hydra* Nematocysts in the Flatworm *Microstomum lineare*," *European Journal of Cell Biology* 97, no. 4 (2018): 289–99, and Bart Tessens, Marlies Monnen, Merlijn Gijbels, and Tom Artois, "A Tool for Feeding and Mating: The Swiss Army Stylet of *Gyratrix hermaphroditus*," May 30, 2023, http://hdl.handle.net/1942/41553.

8. David Strayer, "The Benthic Micrometazoans of Mirror Lake, New Hampshire," *Archiv für Hydrobiologie Supplementband* 72, no. 3 (1985): 287–426.

9. For an example of an attempt to estimate how many animals end up getting eaten by predators, see David L. Strayer, "Perspectives on the Size Structure of the Lacustrine Zoobenthos, Its Causes, and Its Consequences," *Journal of the North American Benthological Society* 10, no. 2 (1991): 210–21.

10. Thomas L. Kieft, "Microbiology of the Deep Continental Biosphere," in *Their World: A Diversity of Microbial Environments*, ed. Christon J. Hurst (Cham: Springer, 2016), 235–36.

11. Kieft, "Microbiology of the Deep Continental Biosphere," 227–32; Cara Magnabosco, Li-Hung Lin, Hailiang Dong, Malin Bomberg, William C. Ghiorse, Helga Stan-Lotter, Karsten Pedersen, et al., "The Biomass and Biodiversity of the Continental Subsurface," *Nature Geoscience* 11, no. 10 (2018): 707–17.

12. Christian Griebler and Tillmann Lueders, "Microbial Biodiversity in Groundwater Ecosystems," *Freshwater Biology* 54, no. 4 (2009): 649–77; Michael Vernarsky, Matthew L. Niemiller, Cene Fišer, Nathanaelle Saclier, Oana Teodora Moldovan, "Life Histories in Groundwater Organisms," in *Groundwater Ecology and Evolution*, ed. Florian Malard, Christian Griebler, and Sylvie Rétaux (London: Academic Press, 2023), 439–56.

13. Birgit Luef, Kyle R. Frischkorn, Kelly C. Wrighton, Hoi-Ying N. Holman, Giovanni Birarda, Brian C. Thomas, Andrea Singh, et al., "Diverse Uncultivated Ultra-Small Bacterial Cells in Groundwater," *Nature Communications* 6 (2015): 6372; Cindy J. Castelle, Christopher T. Brown, Karthik Anantharaman, Alexander J. Probst, Raven H. Huang, and Jillian F. Banfield, "Biosynthetic Capacity, Metabolic Variety and Unusual Biology in the CPR and DPANN Radiations," *Nature Reviews Microbiology* 16, no. 10 (2018): 629–45; Christine He et al., "Genome-Resolved Metagenomics Reveals Site-Specific Diversity of Episymbiotic CPR Bacteria and DPANN Archaea in Groundwater Ecosystems," *Nature Microbiology* 6, no. 3 (2021): 354–65.

14. D. C. Smith, "Why Do So Few Animals Form Endosymbiotic Relationships with Photosynthetic Microbes?," *Philosophical Transactions of the Royal Society*

of London, Series B 333, no. 1267 (1991): 225–30; Jenny Melo Clavijo, Alexander Donath, João Serôdio, and Gregor Christa, "Polymorphic Adaptations in Metazoans to Establish and Maintain Photosymbioses," *Biological Reviews* 93, no. 4 (2018): 2006–20.

15. Amit K. Singh, Sunil Prabhakar, and Sanjay P. Sane, "The Biomechanics of Fast Prey Capture in Aquatic Bladderworts," *Biology Letters* 7, no. 4 (2011): 547–50; Simon Poppinga, Carmen Weisskopf, Anna Sophia Westermeier, Tom Masselter, and Thomas Speck, "Fastest Predators in the Plant Kingdom: Functional Morphology and Biomechanics of Suction Traps Found in the Largest Genus of Carnivorous Plants," *AoB Plants* 8 (2016): plv140.

16. More detail about the metabolic capabilities of bacteria and archaeans and their roles in ecosystems is available in Eugene L. Madsen, *Environmental Microbiology: From Genomes to Biogeochemistry* (Hoboken, NJ: Wiley, 2016), Stuart E. G. Findlay, Stephen K. Hamilton, David Strayer, and Kathleen C. Weathers, "Microbially Mediated Redox Reactions and Their Role in Ecosystems," in *Fundamentals of Ecosystem Science*, 2nd ed., ed. Kathleen C. Weathers, David L. Strayer, and Gene E. Likens (London: Academic Press, 2021), 103–12, Ian D. Jones and John P. Smol, eds., *Wetzel's Limnology: Lake and River Ecosystems*, 4th ed. (London: Academic Press, 2024), and Tom Fenchel and Gary M. King, *Bacterial Biogeochemistry: The Ecophysiology of Mineral Cycling*, 4th ed. (London: Academic Press, 2024).

CHAPTER 12

1. *The Adventures of Ozzie and Harriet* was a popular situation comedy that ran on American television from 1952 to 1966 and that has been widely rerun. The sexual relationships depicted on this show were typical of those represented on the American television of my youth—there weren't any. A peck on the cheek was about as far as it went, and a viewer might reasonably conclude that a woman became pregnant by ironing a man's shirts. And as in most television shows of that era, any implied sex was strictly monogamous and heterosexual, so I use Ozzie and Harriet as a convenient stereotype for that kind of sexual relationship.

2. Sarah P. Otto, "The Evolutionary Enigma of Sex," *American Naturalist* 174 (2009): S1–S14.

3. Mitchell J. Weiss and Dennis P. Levy, "Sperm in 'Parthenogenetic' Freshwater Gastrotrichs," *Science* 205, no. 4403 (1979): 302–3; Margaret R. Hummon, "Reproduction and Sexual Development in a Freshwater Gastrotrich," pt. 4,

"Life History Traits and the Possibility of Sexual Reproduction," *Transactions of the American Microscopical Society* 105 (1986): 97–109.

4. Hummon, "Reproduction and Sexual Development in a Freshwater Gastrotrich."

5. Gerhard Bauer, "Reproductive Strategy of the Freshwater Pearl Mussel *Margaritifera margaritifera*," *Journal of Animal Ecology* 56, no. 2 (1987): 691–704.

6. David H. Funk, Bernard W. Sweeney, and John K. Jackson, "Why Stream Mayflies Can Reproduce Without Males But Remain Bisexual: A Case of Lost Genetic Variation," *Journal of the North American Benthological Society* 29, no. 4 (2010): 1258–66.

7. Olivia P. Judson and Benjamin B. Normark, "Ancient Asexual Scandals," *Trends in Ecology and Evolution* 11, no. 2 (1996): 41–46.

8. Lyubov Shmakova, Stas Malavin, Nataliia Iakovenko, Tatiana Vishnivetskaya, Daniel Shain, Michael Plewka, and Elizaveta Rivkina, "A Living Bdelloid Rotifer from 24,000-Year-Old Arctic Permafrost," *Current Biology* 31, no. 11 (2021): R712–R713.

9. See Paul Simion, Jitendra Narayan, Antoine Houtain, Alessandro Derzelle, Lyam Baudry, Emilien Nicolas, Rohan Arora et al., "Chromosome-Level Genome Assembly Reveals Homologous Chromosomes and Recombination in Asexual Rotifer *Adineta vaga*," *Science Advances* 7, no. 41 (2021): eabg4216, and references cited therein for an introduction to recent studies on bdelloid reproduction.

10. Simion et al., "Chromosome-Level Genome Assembly Reveals Homologous Chromosomes and Recombination in Asexual Rotifer *Adineta vaga*."

11. Eugene A. Gladyshev, Matthew Meselson, and Irina R. Arkhipova, "Massive Horizontal Gene Transfer in Bdelloid Rotifers," *Science* 320, no. 5880 (2008): 1210–13; Chiara Boschetti, Adrian Carr, Alastair Crisp, Isobel Eyres, Yuan Wang-Koh, Esther Lubzens, and Timothy G. Barraclough et al., "Biochemical Diversification Through Foreign Gene Expression in Bdelloid Rotifers," *PLoS Genetics* 8, no. 11 (2012): e1003035; Isobel Eyres, Chiara Boschetti, Alastair Crisp, Thomas P. Smith, Diego Fontaneto, Alan Tunnacliffe, and Timothy G. Barraclough, "Horizontal Gene Transfer in Bdelloid Rotifers is Ancient, Ongoing, and More Frequent in Species from Desiccating Habitats," *BMC Biology* 13 (2015): article 90; Simion et al., "Chromosome-Level Genome Assembly Reveals Homologous Chromosomes and Recombination in Asexual Rotifer *Adineta vaga*."

12. Robin J. Smith, Renate Matzke-Karasz, Takahiro Kamiya, and Patrick De Deckker, "Sperm Lengths of Non-Marine Cypridoidean Ostracods (Crustacea),"

Acta Zoologica 97, no. 1 (2016): 1–17. Some fruit fly fans may argue that *Drosophila* sperm can be *longer* (as a proportion of body length) than ostracod sperm, but *Drosophila* sperm are much thinner than ostracod sperm, so in terms of *mass*, ostracod sperm are larger.

13.   Smith, "Sperm Lengths of Non-Marine Cypridoidean Ostracods (Crustacea)," 1.

14.   Brigitta I. van Tussenbroek, Nora Villamil, Judith Márquez-Guzmán, Ricardo Wong, L. Verónica Monroy-Velázquez, and Vivianne Solis-Weiss, "Experimental Evidence of Pollination in Marine Flowers by Invertebrate Fauna," *Nature Communications* 7, no. 1 (2016): 12980.

15.   G. Evelyn Hutchinson, *A Treatise on Limnology*, 4 vols. (New York: Wiley, 1957– 93), 3:230; Paul A. Cox, "Hydrophilous Pollination," *Annual Review of Ecology and Systematics* 19 (1988): 261–80. Cox notes that the Reverend William Paley regarded the pollination biology of water celery as so marvelous that he used it as a proof of the existence of God in his influential book *Natural Theology: Or, Evidences of the Existence and Attributes of the Deity, Collected from the Appearances of Nature*, originally published in 1802 and reissued many times.

CHAPTER 13

1.   "Great Stink," *Wikipedia,* https://en.wikipedia.org/w/index.php?title=Great _Stink&oldid=1176810691, accessed November 6, 2023.

2.   The causes, effects, and management of eutrophication are discussed in detail in most freshwater ecology textbooks; see, for example, Walter K. Dodds and Matt R. Whiles, *Freshwater Ecology: Concepts and Environmental Applications of Limnology*, 2nd ed. (Burlington, VT: Academic Press, 2010), 469–507, and Ian D. Jones and John P. Smol, eds., *Wetzel's Limnology: Lake and River Ecosystems*, 4th ed. (London: Academic Press, 2024), 394–99, 1052–53.

3.   For a good recent description of pollution in tropical Asian rivers, see David Dudgeon, *Threatened Freshwater Animals of Tropical East Asia* (London: Routledge, 2023), 72–82.

4.   Dudgeon, *Threatened Animals of Tropical East Asia*, 73.

5.   David L. Strayer, *The Hudson Primer: The Ecology of an Iconic River* (Berkeley: University of California Press, 2012), 140–46; US Environmental Protection Agency, "Hudson River PCBs Superfund Site," https://www.epa.gov/hudson riverpcbs, accessed November 6, 2023.

6.   Sujay S. Kaushal, Gene E. Likens, Michael L. Pace, Jenna E. Reimer, Carly M. Maas, Joseph G. Galella, Ryan M. Utz et al., "Freshwater Salinization Syndrome: From Emerging Global Problem to Managing Risks," *Biogeochemistry* 154, no. 2 (2021): 255–92.

7. Charles T. Driscoll, Gregory B. Lawrence, Arthur J. Bulger, Thomas J. Butler, Christopher S. Cronan, Christopher Eagar, and Kathleen F. Lambert et al., "Acidic Deposition in the Northeastern United States: Sources and Inputs, Ecosystem Effects, and Management Strategies," *BioScience* 51, no. 3 (2001): 180–98.

8. Emily S. Bernhardt, Emma J. Rosi, and Mark O. Gessner, "Synthetic Chemicals as Agents of Global Change," *Frontiers in Ecology and the Environment* 15, no. 2 (2017): 84–90.

9. I know this is crude, but it's a common saying in the United States right now, and it certainly applies here.

10. Herbert Güttinger and Werner Stumm, "Ecotoxicology: An Analysis of the Rhine Pollution Caused by the Sandoz Chemical Accident, 1986," *Interdisciplinary Science Reviews* 17, no. 2 (1992): 127–36; Walter Reinhard, "The SANDOZ Catastrophe and the Consequences for the River Rhine," in *Risk Assessment as a Basis for the Forecast and Prevention of Catastrophies [sic]*, edited by Ion Apostol et al. (Amsterdam: IOS Press, 2008), 113–21.

11. "Kalamazoo River oil spill," *Wikipedia*, https://en.wikipedia.org/w/index.php?title=Kalamazoo_River_oil_spill&oldid=1176963182, accessed November 6, 2023.

12. US Fish and Wildlife Service, *Final Restoration Plan and Environmental Assessment for the Certus Chemical Spill Natural Resource Damage Assessment* (Gloucester: US Fish and Wildlife Service Region 5 Virginia Field Office, 2004).

13. Cindy Chaffee, *Joint Environmental Assessment and Restoration Plan for the Fish Creek #2 Diesel Fuel Spill*, U.S. Fish and Wildlife Service, Indiana Department of Environmental Management, Indiana Department of Natural Resources, and Ohio Environmental Protection Agency, February 24, 1997, https://www.cerc.usgs.gov/orda_docs/DocHandler.ashx?task=get&ID=217.

14. International Commission on Large Dams, "World Register of Dams," https://www.icold-cigb.org/GB/world_register/world_register_of_dams.asp, accessed November 8, 2023.

15. J. Jed Brown et al., "Fish and Hydropower on the US Atlantic Coast: Failed Fisheries Policies from Half-Way Technologies," *Conservation Letters* 6, no. 4 (2013): 280–86; John Waldman, *Running Silver: Restoring Atlantic Rivers and Their Great Fish Migrations* (Guilford, CT: Lyons Press, 2013).

16. Charles Lydeard and Richard L. Mayden, "A Diverse and Endangered Aquatic Ecosystem of the Southeast United States," *Conservation Biology* 9, no. 4 (1995): 800–805; George W. Benz and David E. Collins, eds., *Aquatic Fauna in Peril: The Southeastern Perspective* (Decatur, GA: Lenz Design and Communications, 1997).

17.   Wendell R. Haag, *North American Freshwater Mussels: Natural History, Ecology, and Conservation* (Cambridge: Cambridge University Press, 2012), 319–23.

18.   Dozens of dams are planned or under construction in these great biodiverse river basins: K.O. Winemiller, Peter B. McIntrye, Leandro Costello, and Etienne Fluet-Chouinard, "Balancing Hydropower and Biodiversity in the Amazon, Congo, and Mekong," *Science* 351, no. 6269 (2016); 128–29; Alexander S. Flecker, Qinru Shi, Rafael M. Almeida, Héctor Angarita, Jonathan M. Gomes-Selman, Roosevelt García-Villacorta, Suresh A. Sethi et al., "Reducing Adverse Effects of Amazon Hydropower Expansion," *Science* 375, no. 6582 (2022): 753–60; "Hydropower," Mekong River Commission for Sustainable Development, https://www.mrcmekong.org/our-work/topics/hydropower, accessed November 8, 2023; Angelo Carlino, Matthias Wildemeersch, Celray James Chawand, Matteo Giuliani, Sebastian Sterl, Wim Thiery, Ann van Griensven, and Andrea Castelletti, "Declining Cost of Renewables and Climate Change Curb the Need for African Hydropower Expansion," *Science* 381, no. 6658 (2023), eadf5848.

19.   Julian D. Olden and Robert J. Naiman, "Incorporating Thermal Regimes into Environmental Flows: Modifying Dam Operations to Restore Freshwater Ecosystem Integrity," *Freshwater Biology* 55, no. 1 (2010): 86–107.

20.   Jackie R. Heinricher and James B. Layzer, "Reproduction by Individuals of a Nonreproducing Population of *Megalonaias nervosa* (Mollusca: Unionidae) Following Translocation," *American Midland Naturalist* 141, no. 1 (1999): 140–48.

21.   Philip Micklin, "The Aral Sea Disaster," *Annual Review of Earth and Planetary Sciences* 35 (2007): 47–72; Philip Micklin, N.V. Aladin, and Igor Plotnikov, eds., *The Aral Sea: The Devastation and Partial Rehabilitation of a Great Lake* (Berlin: Springer, 2014).

22.   David Dudgeon, *Freshwater Biodiversity: Status, Trends and Conservation* (Cambridge: Cambridge University Press, 2020), 220; Richard Stone, "Saving Iran's Great Salt Lake," *Science* 349, no. 6252 (2015): 1044–47; Masoud Parsinejad, David E. Rosenberg, Yusuf Alizade Govarchin Ghale, Bahram Khazaei, Sarah E. Null, Omid Raja, Ammar Safaie et al., "40-Years of Lake Urmia Restoration Research: Review, Synthesis, and Next Steps," *Science of the Total Environment* 832 (2022): 155055.

23.   Marc F. P. Bierkens and Yoshihide Wada, "Non-Renewable Groundwater Use and Groundwater Depletion: A Review," *Environmental Research Letters* 14, no. 6 (2019): 063002; Bridget R. Scanlon, Sarah Fakhreddine, Ashraf Rateb, Inge de Graaf, Jay Famiglietti, Tom Gleeson, Quentin Grafton et al., "Global Water Resources and the Role of Groundwater in a Resilient Water Future," *Nature*

Reviews Earth and Environment 4, no. 5 (2023): 87–101; Mira Rojanasakul, Christopher Flavelle, Blacki Migliozzi, and Eli Murray, "America is Using Up Its Groundwater Like There's No Tomorrow," *New York Times*, August 28, 2023.

24. I couldn't find precise figures on the numbers and sizes of dams being built or removed, but many dams are still being built or planned, many of which will be large, whereas the number of dams being removed is modest (a few dozen per year in the United States, one of the leaders in dam removals), and most are small. It seems clear that, globally, rivers are being increasingly dammed. See Christiane Zarfl, Alexander Lumsdon, Laura Tydecks, and Jürgen Berlekamp, "A Global Boom in Hydropower Dam Construction," *Aquatic Sciences* 77, no. 1 (2015): 161–70, Jim Best, "Anthropogenic Stresses on the World's Big Rivers," *Nature Geoscience* 12, no. 1 (2019): 7–21, Dudgeon, *Freshwater Biodiversity*, 225–228, American Rivers, *Free Rivers: The State of Dam Removals in the United States*, https://www.americanrivers.org/wp-content/uploads/2023/02 /DamList2021_Report_02172022_FINAL3.pdf, February 2022, and American Rivers, "Dam Removals Continue across the U.S. in 2022," https://www .americanrivers.org/2023/02/dam-removals-continue-across-the-u-s-in-2022.

25. J. David Allan, Robin Abell, Zeb Hogan, Carmen Revenga, Brad W. Taylor, Robin L. Welcomme, and Kirk Winemiller, "Overfishing of Inland Waters," *BioScience* 55, no. 12 (2005): 1041–51.

26. Allan et al., "Overfishing of Inland Waters"; Fengzhi He, Christiane Zarfl, Vanessa Bremerich, Jonathan N. W. David, Zeb Hogan, Gregor Kalinkat, Klement Tockner, and Sonja C. Jähnig, "The Global Decline of Freshwater Megafauna," *Global Change Biology* 25, no. 11 (2019): 3883–92.

27. Dudgeon, *Threatened Animals of Tropical East Asia*, 241–243; Cheryl Claassen, "Washboards, Pigtoes, and Muckets: Historic Musseling in the Mississippi Watershed," *Historical Archaeology* 28, no. 2 (1994): 1–145; James L. Anthony and John A. Downing, "Exploitation Trajectory of a Declining Fauna: A Century of Freshwater Mussel Fisheries in North America," *Canadian Journal of Fisheries and Aquatic Sciences* 58, no. 10 (2001): 2071–90.

28. FAO, *The State of World Fisheries and Aquaculture 2022: Towards Blue Transformation* (Rome: FAO, 2022), https://doi.org/10.4060/cc0461en.

29. Hanno Seebens, Laura A. Meyerson, Sebataolo J. Rahlao, Bernd Lenzner, Elena Tricarico, Alla Aleksanyan, Franck Courchamp et al., "Trends and Status of Alien and Invasive Alien Species," in *Thematic Assessment Report on Invasive Alien Species and Their Control of the Intergovernmental Science-Policy Platform on Biodiversity and Ecosystem Services*, ed. Helen E. Roy, Aníbal Pauchard, Peter Stoett, and Tanara Renard Truong (Bonn: IPBES Secretariat, 2023).

30.   David L. Strayer, "Alien Species in Fresh Waters: Ecological Effects, Interactions with Other Stressors, and Prospects for the Future," *Freshwater Biology* 55, no. s1 (2010): 152–74.

31.   David L. Strayer et al., "Decadal-Scale Change in a Large-River Ecosystem," *BioScience* 64, no. 1 (2014): 496–510, and references cited therein.

32.   Many papers show rising temperatures, decreasing periods of ice cover, and so forth in inland waters, for example, R. Iestyn Woolway, Benjamin M. Kraemer, John D. Lenters, Christopher J. Merchant, Catherine M. O'Reilly, and Sapna Sharma, "Global Lake Responses to Climate Change," *Nature Reviews Earth and Environment* 1, no. 8 (2020): 388–403.

33.   Donovan A. Bell, Ryan P. Kovach, Clint C. Muhlfeld, Robert Al-Chokhachy, Timothy J. Cline, Diane C. Whited, David A. Schmetterling et al., "Climate Change and Expanding Invasive Species Drive Widespread Declines of Native Trout in the Northern Rocky Mountains, USA," *Science Advances* 7, no. 52 (2021): eabj5471.

34.   William G. McDowell, Amy J. Benson, and James E. Byers, "Climate Controls the Distribution of a Widespread Invasive Species: Implications for Future Range Expansion," *Freshwater Biology* 59, no. 4 (2014): 847–57; Mafalda Gama, Daniel Crespo, Marina Dolbeth, and Pedro Manuel Anastácio, "Ensemble Forecasting of *Corbicula fluminea* Worldwide Distribution: Projections of the Impact of Climate Change," *Aquatic Conservation* 27, no. 3 (2017): 675–84.

35.   See, for example, Sami Domisch, Miguel B. Araújo, Núria Bonada, Steffen U. Pauls, Sonja C. Jähnig, and Peter Haase, "Modelling Distribution of European Stream Macroinvertebrates under Future Climates," *Global Change Biology* 19, no. 3 (2013): 752–62; Katrina L. Pound, Chad A. Larson, and Sophia I. Passy, "Current Distributions and Future Climate-Driven Changes in Diatoms, Insects and Fish in U.S. Streams," *Global Ecology and Biogeography* 30, no. 1 (2021): 63–78; Amy R. Tims and Erin E. Saupe, "Forecasting Climate-Driven Habitat Changes for Australian Freshwater Fishes," *Diversity and Distributions* 29, no. 5 (2023): 641–53.

36.   If you want more information about climate change, the IPCC (www.ipcc .ch) is an excellent source of scientifically sound information. Its "Summary for Policymakers" (https://www.ipcc.ch/report/ar6/syr/downloads/report /IPCC_AR6_SYR_SPM.pdf) is a good readable overview, and its full reports contain all of the details.

37.   "South–North Water Transfer Project," *Wikipedia*, https://en.wikipedia .org/w/index.php?title=South%E2%80%93North_Water_Transfer_Project& oldid=1179088764, accessed November 12, 2023.

38.   Matthew R. Fuller, Martin W. Doyle, and David L. Strayer, "Causes and Consequences of Habitat Fragmentation in River Networks," *Annals of the New York Academy of Sciences (The Year in Ecology and Conservation Biology)* 1355, no. 1 (2015): 31–51.

39.   For an introduction, see J. David Allan, "Landscapes and Riverscapes: The Influence of Land Use on Stream Ecosystems," *Annual Review of Ecology, Evolution, and Systematics* 35 (2004): 257–84.

40.   Milton B. Trautman, *The Fishes of Ohio*, revised ed. (Columbus: Ohio State University Press, 1981), 13–35, 446.

41.   David L. Strayer and David Dudgeon, "Freshwater Biodiversity Conservation: Recent Progress and Future Challenges," *Journal of the North American Benthological Society* 29, no. 1 (2010): 344–58; Bernhardt et al., "Synthetic Chemicals as Agents of Global Change."

42.   Dudgeon, *Freshwater Biodiversity*.

43.   David L. Strayer, "Challenges for Freshwater Invertebrate Conservation," *Journal of the North American Benthological Society* 25, no. 2 (2006): 271–87; Strayer and Dudgeon, "Freshwater Biodiversity Conservation."

44.   B. D. Smith, Ding Wang, Gill T. Braulik, Randall Reeves, Kaiya Zhou, Jay Barlow, and Robert L. Pitman, "*Lipotes vexillifer*," *IUCN Red List of Threatened Species* (2017): e.T12119A50362206. https://dx.doi.org/10.2305/IUCN.UK.2017-3.RLTS. T12119A50362206.en; Dudgeon, *Threatened Animals of Tropical East Asia*, 306.

45.   Hui Zhang, Ivan Jarić, David L. Roberts, Yongfeng He, Hao Du, Jinming Wu, Chengyou Wang et al., "Extinction of One of the World's Largest Freshwater Fishes: Lessons for Conserving the Endangered Yangtze Fauna," *Science of the Total Environment* 710 (2020): 136242.

46.   Dudgeon, *Threatened Animals of Tropical East Asia*, 251–52.

47.   NatureServe, "NatureServe Explorer," https://explorer.natureserve.org; James D. Williams, Arthur E. Bogan, Robert S. Butler, Kevin S. Cummings, Jeffrey T. Garner, John L. Harris, Nathan A. Johnson et al., "A Revised List of the Freshwater Mussels (Mollusca: Bivalvia: Unionida) of the United States and Canada," *Freshwater Mollusk Biology and Conservation* 20, no. 2 (2017): 33–58.

48.   Trautman, *The Fishes of Ohio*, 611–13; Amanda E. Haponski and Carol A. Stepien, "A Population Genetic Window into the Past and Future of the Walleye *Sander vitreus*: Relation to Historic Walleye and the Extinct 'Blue Pike' *S. v. 'glaucus,'*" *BMC Evolutionary Biology* 14 (2014): article 133. Landings data come from Great Lakes Fishery Commission, *Commercial Fish Production in the Great Lakes 1867–2020* (Ann Arbor, MI: Great Lakes Fishery Commission, 2022), www.glfc.org/great-lakes-databases.php.

CHAPTER 14

1. Beth Shapiro's excellent book *How to Clone a Mammoth* (Princeton, NJ: Princeton University Press, 2015) describes both the potential and the limitations of de-extinction and explains the technical details, including why it is infeasible to revive or even produce close facsimiles of many extinct species. Many ecologists are wary about the costs and risks of de-extinction, which in any case certainly will not fix the problems of human-caused extinctions; see Paul Ehrlich and Anne H. Ehrlich, "The Case Against De-Extinction: It's a Fascinating But Dumb Idea," *Yale Environment 360*, January 13, 2014, https://e360.yale.edu/features/the_case_against_de-extinction_its_a_fascinating_but_dumb_idea, and Dave Strayer, "De-Extinction, a Risky Ecological Experiment," *Ecotone*, February 19, 2016, https://www.esa.org/esablog/2016/02/19/de-extinction-a-risky-ecological-experiment/.

2. Sometimes it is possible to control well-established invaders, as detailed in digression 14.1. But this requires the right combination of money, tolerance for harmful side effects, and luck and is unlikely to be feasible for all (or even a majority of) established invaders. Critics who dismiss biological invasions as a serious problem sometimes say that development of gene drives will soon give us a silver bullet to get rid of harmful invaders. However, before we can deploy gene drives to control invaders, we will need to solve serious technical and ethical problems, and I do not think that gene drives or anything else is likely to provide safe, effective, and economical control of most established invaders anytime soon. See, for example, Kenneth A. Oye, Kevin Esvelt, Evan Appleton, Flaminia Catteruccia, George Church, Todd Kuiken, Shlomiya Bar-Yam Lightfoot et al., "Regulating Gene Drives," *Science* 345, no. 6197 (2014): 626–28; Kent H. Redford, Nicholas B. W. Mcfarlane, Thomas Brooks, and Jonathan S. Adams., eds., *Genetic Frontiers for Conservation: An Assessment of Synthetic Biology and Biodiversity Conservation* (Gland: International Union for the Conservation of Nature, 2019); John L. Teem, Luke Alphey, Sarah Descamps, Matt P. Edgington, Owain Edwards, Neil Gemmell, Tim Harvey-Samuel et al., "Genetic Biocontrol for Invasive Species," *Frontiers in Bioengineering and Biotechnology* 8 (2020): 00452.

3. If you're interested in more detail about solutions, look at Rebecca Flitcroft, Michael S. Cooperman, Ian J. Harrison, Diego Juffe-Bignoli, and Philip J. Boon, "Theory and Practice to Conserve Freshwater Biodiversity in the Anthropocene," *Aquatic Conservation* 29, no. 7 (2019): 1013–21, David Dudgeon, *Freshwater Biodiversity: Status, Trends and Conservation* (Cambridge: Cambridge University

Press, 2020), and David Tickner, Jeffrey J. Opperman, Robin Abell, Mike Acre-man, Angela H. Arthington, Stuart E. Bunn, and Steven J. Cooke et al., "Bending the Curve of Global Freshwater Biodiversity Loss: An Emergency Recovery Plan," *BioScience* 70, no. 4 (2020): 330–42, and references cited therein.

4.   Liuyong Ding, Liqiang Chen, Chengzhi Ding, and Juan Tao, "Global Trends in Dam Removal and Related Research: A Systematic Review Based on Associ-ated Datasets and Bibliometric Analysis," *Chinese Geographical Science* 29, no. 1 (2019): 1–12; "Dam Removal," *Wikipedia* https://en.wikipedia.org/w/index .php?title=Dam_removal&oldid=1177082614, accessed November 9, 2023.

5.   Angela H. Arthington, *Environmental Flows: Saving Rivers in the Third Millen-nium* (Berkeley: University of California Press, 2012).

6.   To see how this might be done, see Alexander S. Flecker, Qinru Shi, Rafael M. Almeida, Héctor Angarita, Jonathan M. Gomes-Selman, Roosevelt García-Villacorta, Suresh A. Sethi et al., "Reducing Adverse Effects of Amazon Hy-dropower Expansion," *Science* 375, no. 6582 (2022): 753–60.

7.   New York City Department of Environmental Protection, "Historical Drought & Water Consumption Data," https://www.nyc.gov/site/dep/water/histo ry-of-drought-water-consumption.page, accessed November 3, 2022.

8.   Margaret A. Palmer, Joy B. Zedler, and Donald A. Falk, eds., *Foundations of Restoration Ecology*, 2nd ed. (Washington, DC: Island Press, 2016); Karen D. Holl, *Primer of Ecological Restoration* (Washington, DC: Island Press, 2020).

9.   See, for example, Margaret A. Palmer, Holly L. Menninger, and Emily Bern-hardt, "River Restoration, Habitat Heterogeneity and Biodiversity: A Failure of Theory or Practice?," *Freshwater Biology* 55, no. s1 (2010): 205–22, Katherine N. Suding, "Toward an Era of Restoration in Ecology: Successes, Failures, and Opportunities Ahead," *Annual Review of Ecology, Evolution, and Systematics* 42 (2011): 465–87, and Kyle Van den Bosch and Jeffrey W. Matthews, "An Assess-ment of Long-term Compliance with Performance Standards in Compensatory Mitigation Wetlands," *Environmental Management* 59, no. 4 (2017): 546–56.

10. For a discussion of the potential and difficulties of eradicating invasive spe-cies that includes examples, see Judith H. Myers, Daniel Simberloff, Armand M. Kuris, and James R. Carey, "Eradication Revisited: Dealing with Exotic Species," *Trends in Ecology and Evolution* 15, no. 8 (2000): 316–20, and Dan-iel Simberloff, "Maintenance Management and Eradication of Established Aquatic Invaders," *Hydrobiologia* 848, no. 9 (2021): 2399–420.

11. Rochelle A. Sturtevant, Dora M. Mason, Edward S. Rutherford, Ashley Elgin, El Lower, and Felix Martinez, "Recent History of Nonindigenous Species in

the Laurentian Great Lakes; An Update to Mills et al., 1993 (25 Years Later)," *Journal of Great Lakes Research* 45, no. 6 (2019): 1011–35.

12.   A good account of sea lamprey and its control in the Great Lakes is provided by Cory Brant, *Great Lakes Sea Lamprey: The 70 Year War on a Biological Invader* (Ann Arbor: University of Michigan Press, 2019).

13.   Great Lakes Fishery Commission, *Commercial Fish Production in the Great Lakes, 1867–2020* (Ann Arbor, MI: Great Lakes Fishery Commission, 2022), www.glfc.org/great-lakes-databases.php.

14.   Keith R. Edwards, "*Lythrum salicaria* L. (purple loosestrife)," in *A Handbook of Global Freshwater Invasive Species*, ed. Robert A. Francis (New York: Routledge, 2012), 91–102; Stacy B. Endriss, Victoria Nuzzo, and Bernd Blossey, "Success Takes Time: History and Current Status of Biological Control of Purple Loosestrife in the United States," in *Contributions of Classical Biological Control to the U.S. Food Security, Forestry, and Biodiversity*, ed. Roy G. Van Driesche, Rachel L. Winston, Thomas M. Perring, and Vanessa M. Lopez (Morgantown, WV: USDA Forest Service, 2022), 312–28.

15.   Ballast water is fresh or salt water taken on by a ship to add weight and improve its stability and maneuverability, especially if it isn't carrying a full load of cargo or is traveling in rough seas. The ballast holds of large ocean-going ships are huge and may contain tens of millions of gallons (hundreds of thousands of cubic meters) of water. Until recently, ballast water received little or no treatment, so it was full of living organisms, from viruses and bacteria to fish, and was responsible for a great many invasions of troublesome species around the world, in both fresh and salt water (think of ballast holds as enormous, floating aquaria). Ballast water and its inhabitants often were released into the environment when ships took on cargo in new ports. The international community has been moving to treat or eliminate ballast water to cut down on invasions, but large volumes of untreated or incompletely treated ballast water are still moving around the world.

16.   Sigal Samuel, "Lake Erie Has Legal Rights, Just Like You," *Vox*, February 26, 2019, https://www.vox.com/future-perfect/2019/2/26/18241904/lake-erie-legal -rights-personhood-nature-environment-toledo-ohio; Jack Zouhary, "Order Invalidating Lake Erie Bill of Rights," February 27, 2020, https://www.ohio attorneygeneral.gov/Files/Briefing-Room/News-Releases/Environmental -Enforcement/2020-02-27-Decision-invalidating-LEBOR.aspx.

INDEX

Page numbers in *italics* refer to illustrations; the letter t following a page number denotes a table.

shads, 83, 137

sharks, 64

Sierra Club, 167

sinkholes, 10, 30, 35

snails: conservation status of, 149–50;
extinctions of, 138; migrations of, 84

snakes, 66, 108

South-North Water Transfer Project,
146

sperm, giant, 126–27

spills, of pollutants, 135–36

spiny-headed worms, 79

sponges, 39, 72, 73; photosynthesis in,
115

stoneflies, 109; conservation status of,
149–50

stoneworts, 75

stormwater, 17, 31

stream restoration. *See* ecological
restoration

streams and rivers, 11–13; ages of,
37–38; ancient, 38, 40; ephemeral,
44, 95–104; largest, 11–12, 19, 20t;
numbers of, 5, 12; oldest, 37–38;
origins of, 31–33; types of, 11–13, 32;
underground, 32

sturgeons, 65, 83, 137, 143, 152–53

sundews, 115

swamps. *See* wetlands

synthetic biology, 42

Syr Darya sturgeon, 65

Tahoe, Lake, 30

Tanganyika, Lake, 20t, 30; unique
species in, 39

tapeworms, 80

tardigrades. *See* water bears

Tasmanian crayfish, 72–73

Tennessee River, 40, 138

Thames River, 133

theme and variations, 3–5, 169

threats to biodiversity, 131–49

Titicaca, Lake, unique species in, 39

Torrens, Lake, 44

toxic algae, 75, 76, 77, 133, 135

"tragedy of the commons," 143

Trout Unlimited, 166

Tulare Lake, 14–15

turtles, 66, 67; imperilment of, 152;
overharvest of, 143

Urmia, Lake, 142

values of inland waters, 2, 132, 171;
competing, 132

Venus flytraps, 115

vernal pools, 44, 104

Victoria (water lily), 74

Victoria, Lake, 6, 20t, 30

volcanoes, 33, 35, 55; as creators of
lakes, 29–30

Vostock, Lake, 19, 20t

water bears/tardigrades: diet of, 106;
dormancy in, 97, 99–101; in outer
space, 99–100; remarkable environmental tolerances of, 99

water chemistry, 50–57; influence of
watershed on, 54–56

water clarity, 57, 144; clearest lake in
the world, 184n8

water conservation, 158–59, 165

water diversions, 139, 146

waterfalls, 32

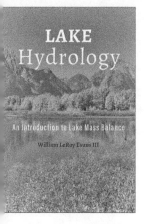
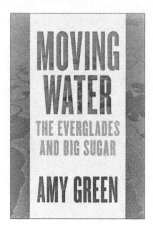
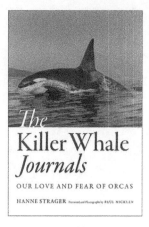
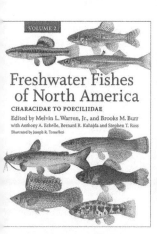
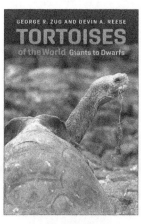
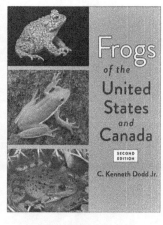